Kant and Technics

Also Available from Bloomsbury

The Kant Dictionary, Lucas Thorpe
The Bloomsbury Companion to Kant, ed. by Gary Banham, Dennis Schulting, and Nigel Hems
Kant's Humorous Writings, Robert R. Clewis

Kant and Technics

From the Critique of Pure Reason *to the* Opus Postumum

Franziska Aigner

BLOOMSBURY ACADEMIC
LONDON • NEW YORK • OXFORD • NEW DELHI • SYDNEY

BLOOMSBURY ACADEMIC
Bloomsbury Publishing Plc
50 Bedford Square, London, WC1B 3DP, UK
1385 Broadway, New York, NY 10018, USA
29 Earlsfort Terrace, Dublin 2, Ireland

BLOOMSBURY, BLOOMSBURY ACADEMIC and the Diana logo are trademarks of Bloomsbury Publishing Plc

First published in Great Britain 2024

Cover design by Charlotte Daniels
Cover image: Kant, Immanuel 22.4.1724- 12.2.1804, German philosopher, works, "Zum ewigen Frieden", 1795, part of the last pages.
Contributor: INTERFOTO / Alamy Stock Photo

A catalogue record for this book is available from the British Library.

A catalog record for this book is available from the Library of Congress.

Library of Congress Cataloging-in-Publication Data

Names: Berger-Soraruff, Amélie, author.
Title: Technics of existence : Sartre, Foucault and Stiegler / Amelie Berger-Soraruff.
Description: London; New York: Bloomsbury Academic, 2025. |
Includes bibliographical references.
Identifiers: LCCN 2024020724 (print) | LCCN 2024020725 (ebook) |
ISBN 9781350416178 (hb) | ISBN 9781350416215 (pb) | ISBN 9781350416192 (epub) |
ISBN 9781350416185 (ebook)
Subjects: LCSH: Stiegler, Bernard. | Sartre, Jean-Paul, 1905-1980. | Foucault, Michel, 1926-1984. | Technology–Philosophy. | Self (Philosophy)
Classification: LCC B2430.S7524 B47 2025 (print) | LCC B2430.S7524 (ebook) |
DDC 194–dc23/eng/20240807
LC record available at https://lccn.loc.gov/2024020724
LC ebook record available at https://lccn.loc.gov/2024020725

ISBN: HB: 978-1-3502-9903-0
ePDF: 978-1-3502-9904-7
eBook: 978-1-3502-9905-4

Typeset by Deanta Global Publishing Services Chennai India

To find out more about our authors and books visit www.bloomsbury.com and sign up for our newsletters.

For my grandmother, Gertrude Fischer, whose love and defiance have always been my guide.

Contents

Acknowledgements

I am fortunate to have written this book in the company of many friends and colleagues. Most of all, I would like to thank my former PhD supervisor, Howard J. Caygill, for having first sparked my interest on technics in Kant, for teaching me how to read Kant and Simondon long before becoming my supervisor and for having become my friend in the process. In one of our first meetings, Howard put me on the trail of an old, blue and long out-of-print book that one might be able to find in the British Library. Written by Gerhard Lehmann, this book would soon become the key to my project in the years to come. Little did I know, back then, of Howard's gift to intuit and provide, so generously, the horizon upon which my philosophical intuitions could develop. I would like to thank Peter Hallward for being the critical, perceptive and sharp reader that he is, and for his continued insistence on rigorous argumentation and clarity. I am greatly indebted to the incisive and critical questions by Peter Osborne, many of which have forced me into crisis, resulting in necessary decisions long overdue. It has been an honour to discuss my project with Étienne Balibar at various moments. His brilliant insights have helped me navigate through difficult passages more than once, and his teachings will stay with me through all that lies ahead. I would also like to thank Stephen Howard for having read my work on the notoriously difficult manuscript of the *Opus Postumum* at various stages throughout the process and for his generous comments, which have both improved my argument and my confidence. Most of all, however, I could not have written this dissertation without the singular intellectual community I found at the Centre for Research in Modern European Philosophy at Kingston University, London. I would like to thank in particular Isabell Dahms, Marie Louise Krogh, Austin Gross, Francis Cooper, Matt Hare, Kylie Gilchrist, Charlotte Szász, Joel White and Nathaniel Wooding for their friendship, their philosophical insights and for having read and discussed my writing with me throughout the years. My immense gratitude also goes to Iain Campbell, who has helped me improve my writing in the final stages. I am also greatly indebted to Lawrence and to my friends at 113 in London, Salzburg and Berlin. My infinite gratitude goes to Sam, who has supported me in all imaginable ways in thought and love.

Versions of Chapter 1 (sections 1 and 3) have been published as 'Confronting the Technical Aporia: Heidegger's and Stiegler's Technics-Thinking' in *Technophany, A Journal for Philosophy and Technology Issue #2:* 0-29.

Guide to Abbreviations

Immanuel Kant's writings are cited in reference to

AA	*Immanuel Kants gesammelte Schriften. 29. Vols.* Herausgegeben von der Preussischen Akademie der Wissenschaften (I–XXII), der deutschen Akademie der Wissenschaften zu Berlin (XXIII) und der Akademie der Wissenschaften zu Göttingen (XXIV–XXIX) (Abteilung I: Werke: I–IX; Abteilung II: Briefwechsel: X–XIII; Abteilung III: Handschriftlicher Nachlass: XIV–XXIII; Abteilung IV: Vorlesungen: XXIV–XXIX) Berlin: 1902.
ANTH	*Anthropology from a Pragmatic Point of View* (AA 7)
CPR	*Critique of Pure Reason* (A-edition: 1781/B-edition: 1787)
CJ	*Critique of the Power of Judgment* (AA 5)
GMM	*Groundwork of the Metaphysics of Morals* (AA 4)
OP	*Opus Postumum* (AA 21–22)

All English translations of Kant's works are to be found in *The Cambridge Edition of the Works of Immanuel Kant* (Cambridge, UK), unless stated otherwise.

Preface

In his programmatic 'First Introduction' to the *Critique of the Power of Judgement*, Kant refers to a 'technical' power of judgement that accords with a 'technique of nature'. What makes this reference so enigmatic is not only Kant's assumption that his readers will be familiar with what he means by technics but also that they would be ready to embark on the painstaking 'search for a principle of the technical power of judgement' – few were of course preferring to look for other things in the third critique. The principle of the technical power of judgement is certainly hard to grasp: it exceeds the division of philosophy into theoretical and practical parts while also being contained within the critique of the faculty of cognition. It is none other than the 'reflecting power of judgement' analysed throughout the third critique but while described as 'technical' does not function mechanically 'like a machine' but rather *artistically or artificially*. The confident yet fugitive programmatic statements of a technical reason beyond theoretical and practical reason still barely prepare for the explosion of the vocabulary of technics that reverberates throughout the third critique rendering it the most elusive and intractable of the critical trilogy.

Fortunately, we now have Franziska Aigner's book that patiently sounds the archipelago of technical reason underlying its dramatic manifestations in the technics of nature and the power of judgement. She patiently reconstructs Kant's philosophy of technics tracing it to its sources and revealing its inconspicuous service as the higher condition of possibility for the entire critical philosophy. She also traces its insistent emergence in Kant's post-critical philosophy and its promotion to the central place in his late philosophy of transition in the *Opus Postumum*.

Her book performs the immense service of making finally legible the thought of technics in Kant's philosophy and in speculating on how this legibility changes much in our understanding of the philosophy of technics. She shows how far the thinking of technics in Heidegger, Simondon and Stiegler depends on an assumption that the technical remains unthought in Kant. Her diagnosis of such occlusion of the technical in Kant's thought as ultimately justifying the existence – or lamented disappearance – of a pure human reason is challenging and must send us back to reconsider the twentieth-century dystopias of technical reason and its civilization.

Beyond its evident scholarly contribution to the history of the philosophy of technics, Aigner's book implicitly addresses contemporary debates and public anxieties around the threats and promises of artificial intelligence. In her reading of Kant, intelligence was always artificial, reason always-already technically invested. Her return to Kant upsets any generic distinction between human and artificial intelligence pointing instead to the existence of different modes of technical reason. This introduces welcome nuance into contemporary ethical and regulatory debates about AI by pointing to a continuum of technically invested reason rather than assuming that human reason was ever not technical. In all its detailed and scrupulous historical scholarship, *Kant and Technics* has the potential to change the way we think and the ways in which we act.

Howard Caygill

Introduction

Kant and/without technics

In 1936, the German *Akademie der Wissenschaften* published the first part of the *Opus Postumum* (*Nachlasswerk*), containing the hitherto unpublished manuscript on which Immanuel Kant had worked over the last decade of his life. In Fascicle I, which holds the chronologically last pages of the manuscript, Kant sketched out the following chapter outline, which bears witness to the systematic importance granted to technical-practical reason (*technisch-praktische Vernunft*) in the *Opus Postumum*:

> Titlesheet and Preface
> The world as universum
>
> In all these objects a maximum idea, *ergo unicum* in all three cases
>
> 1. theoretical-speculative [reason]
> 2. technical-practical [reason]
> 3. moral-practical reason
>
> From intuitions, *a priori* concepts and ideas
> The idea of freedom leads through the categorical imperative to God
>
> 1. speculative [reason]
> 2. practical [reason]
> 3. technical-practical [reason]
> 4. moral-practical reason in one system.[1]

At the very end of his life, Kant had evidently planned to systematically treat technical-practical reason on a par with theoretical, practical and moral-practical reason, and thus as an important part of his system of transcendental philosophy. Differently to theoretical and pure practical reason, however, technical-practical reason had hitherto not been subjected to a critique. Following the first *Critique of Pure Reason* and the second *Critique of Practical Reason* Kant had, of course, written a third critique, tasked with the important function of bridging theoretical and practical reason. This third critique had, however, not been directed to reason, as it instead presented itself as a critique of the power of judgment. As a problem of empirical philosophy, technical-practical reason had had no need of a preliminary critique.

As the chapter outline shows, the *Opus Postumum* placed technical-practical reason in between theoretical and practical reason proper, indicating that a decade after the 1790 *Critique of the Power of Judgment*, it is now technical-practical reason which has taken over the systematic role of bridging theoretical and practical reason. This led the German philosopher and editor of the *Opus Postumum*, Gerhard Lehmann, to the conclusion that the *Opus Postumum* contained the outline to what would have become Kant's third critique of reason had he lived to complete it, only this time addressed to reason in its technical-practical use.[2] On a meta-level, then, technical-practical reason not only bridges theoretical and practical reason but furthermore provides the ground to conjoin the two relata of the title of this book. The *Opus Postumum* contains Kant's late thought on technics.

In his 1954 seminal essay *The Question Concerning Technology*,[3] and thus almost two centuries after the beginning of the Industrial Revolution, Martin Heidegger argued that the right question concerning technics had still not been asked. Addressing both philosophy and culture at large, Heidegger was not the only philosopher who, in the years immediately following the devastating events of the Second World War, problematized a fateful forgetting of technics. In his 1958 book *On the Mode of Existence of Technical Objects*,[4] Gilbert Simondon claimed that technics must be understood as philosophy's aporia, meaning that philosophy is structurally incapable of seeing, let alone addressing, technics. Yet further to this, and in a similar vein to Heidegger, Simondon also located this problem in culture as a whole, which, he wrote, 'has constituted itself as a defence system against technics'.[5] Following both Heidegger and Simondon, Bernard Stiegler's first installment of his *Technics and Time* series was first published in France in 1994. Repeating Heidegger and Simondon's criticisms forty years later, Stiegler argues, in a quasi-Derridean fashion, that the history of philosophy must ultimately be read as the history of the repression of technics by philosophy.[6]

According to all three thinkers, philosophy's technical aporia reaches its clearest expression in one of modernity's most prominent philosophers: Immanuel Kant. According to Heidegger, Kant articulated the essence of modern technics as enframing (*Gestell*) under the name of his *Transcendental Analytic* without understanding the nature of his own insight, while Simondon claimed that technics is the one question that transcendental philosophy is incapable of answering since its answer will either come too early (a priori) or too late (a posteriori). Stiegler synthesizes both positions in his claim that Kant was incapable of seeing the technical constitution of his own consciousness. What becomes clear, then, is that all three thinkers argue, in their own way, that Kant was essentially incapable of seeing, understanding, let alone thinking, technics.

Each of these diagnoses is underlined by the striking absence of references or discussions of Kant's own explicit writing on technics, as if he had had nothing to say on the matter. To say with Heidegger, Simondon, and Stiegler that Kant does not speak of technics thus means that, ultimately, Kant does not have a future beyond the status of the symptom granted. This is the case because philosophy's task in the present conjunction, all three thinkers claim, has become to *think* technics. According to the three self-proclaimed thinkers of technics, technics is philosophy's symptom and Kant's transcendental philosophy has become the symptom of a philosophy that could not think technics. Thus, what in the first paragraph initially appeared as a promising conjunction, inquiring into the relation between Kant and technics, is forcefully undercut by the history of philosophy told by Heidegger, Simondon, and Stiegler. Rather than going by the name of *Kant and technics*, should the title of this book rather read *Kant without technics*?

The problem and the promise

At the centre of this book lies an essentially twofold path. In what follows, I will argue against Heidegger, Simondon, and Stiegler that there is indeed an explicit concept of technics at work in Kant's philosophy. As the following discussion will show, this concept is not *a* technics that was overlooked by Heidegger, Simondon, and Stiegler, but it is rather the case that transcendental philosophy must be understood as *of* technics. As a philosophy *of* technics, transcendental philosophy does not topically engage with technics as a possible object of inquiry among others. Rather, transcendental philosophy is at once constituted against, while at the same time relying on, being built upon, and proceeding from technics. Furthermore, I will show in what follows that the conjunction *and*, correlating the two relata in the title of this book, is subject to development over time, as its meaning changes throughout the trajectory of Kant's thought from the *Critique of Pure Reason* (1781) until the *Opus Postumum* (1796–1803).

On the other hand, however, and in asking about the relation between Kant and technics, this book also enquires into the wider relation between philosophy and technics. Heidegger, Simondon, and Stiegler all claim that philosophy has a problematic relation to technics in the sense that it is structurally incapable of seeing, recognizing, or remembering its own, essentially technical, constitution. Each in their own way can be seen to put forward their respective thought on technics in contradistinction to the history of philosophy through the act of questioning (Heidegger), problematizing (Simondon), or remembering, in the sense of inventing, technical thought anew (Stiegler). Their conceptual arguments are supported and warranted by their reading of the history of philosophy as the history of philosophy's technical oblivion. The methodological challenge faced by each of the three thinkers is great, since they must all, in one way or another, be able to ask, problematize or remember the hitherto structurally impossible. While the singularity of their own position, both epistemologically and methodologically, is a problem that each of them must address, the methodological role played by the history of philosophy is clear. The history of philosophy must harbour the resources for thinking or remembering technics, but only ever implicitly.

However, if there is indeed such a thing as a technical thought proper in Kant, and if that technical thought falls outside what has here been presented as a simple two-form schema, then Kant can no longer play the role of philosophy's most prominent techno-oblivious thinker, and the question of the relation between philosophy and technics as a whole will need to be reopened; the history of philosophy will need to be reinvestigated from the point of view of possible other technics, as there might be, despite Heidegger, Simondon, and Stiegler's inability to see them, explicit philosophical resources for thinking technics proper in the history of philosophy. At stake in the restricted question about the relation between Kant and technics is thus ultimately the relationship between philosophy and technics as a whole.

The promise of the opening encounter with Kant's projected chapter outline in his late *Opus Postumum* is that there is indeed a positive and explicit notion of technics at play in Kant's thinking. As I will argue, this notion is at once developed out of the specifically German context surrounding Kant, responding to the medieval scholastic German notion of *Technologie* as well as to Johann Beckmann's newly founded science of *Technologie*, and at the same time engaging with the Aristotelian notion of *techne*. At the same time, however, Kant's thought on technics also harbours resources for thinking technics from a higher standpoint, as technical-practical reason systematically charges itself with the technical-practical task of the world. Technical-practical reason would thus have the role of forming the cosmo-technical context from which specific notions of technics can then, in a second step, derive their specific meaning and position.

Outline

The following discussion is organized according to four chapters. The first chapter will set the scene by laying out in more detail the field and history of technical thought, according to Heidegger, Simondon, and Stiegler. The aim of this chapter is to shine a light on the details of the paradoxical role occupied by Kant and transcendental philosophy within both the field of technical thought and its history, as presented in Heidegger, Simondon, and Stiegler's respective accounts. In this regard, each thinker will be discussed first on their own concerning their technical thinking, second via their reading of the history of philosophy and finally through Kant's role within both their thinking and their projected history of technical thought.

In Chapter 2, I will leave behind Heidegger, Simondon, and Stiegler's writings in order to begin the task of reading Kant on his own terms. At first glance, Kant's world of universal and necessary forms of thought – the transcendental – surely seems removed from the world of technics, with its problems of technical production and instrumentality. From a methodological perspective, however, I will argue that a technics of instrumentality is the implicit problem that Kant ascribes to reason, warranting the *Critique of Pure Reason* as a whole. According to Kant, reason strives towards speculative expansion by means of the instrument of logic. The explicit aim of the *Critique of Pure Reason* is thus to counteract reason's instrumental transgressions by way of a twofold critical programme. First, Kant will venture to delineate what theoretical reason wants, drives, or strives to know in relation to what it 'can' know, meaning that reason's ends and questions are to be set in relation to its own proper powers, its faculties and forces under the name of the canon. Following the canon laid out in the *Transcendental Doctrine of Elements*, the subsequent *Transcendental Doctrine of Method* continues by asking what reason 'can do' with these materials. I will show that at this point and according to Kant's lexicon, questions of method constitute the, properly speaking, 'technical' part of logic. Following the work of German linguist Wilfried Seibicke, I will argue that Kant employed the term technics (*die Technik*) in line with the medieval scholastic notion of *Technologie*, which he shortened to *die Technik*, in order to refer to methodological problems of terminology and the systematization of terms. At the same time, however, I will argue that this technical doctrine of logic will also take recourse to a second genealogy of technics under the name of a discipline of reason. Only after reason has been disciplined and its skill and talent (its *techne*) have been worked upon will reason want what it can have, as its ends will now have been set in accordance with what it can do.

While Kant's explicit discussion of technics takes place in regard to method, Chapter 2 will also show that both Heidegger and Stiegler's symptomatic readings of technics in Kant consider the *Transcendental Analytic*, and thus the canon, as the place of Kant's implicit technics. This chapter will show that and how Heidegger and Stiegler were not the first to have read the canon put forward by Kant as containing a technical problematic. In a scholarly discussion between Kant and his contemporary J. G. Schlosser,[7] Kant was already accused during his own lifetime of having reduced cognition to 'a mere form-giving manufacture [*Formgebungs-Manufactur*]'.[8] Despite Kant's outright and polemic dismissal of the critique, I will show that Kant insists on the distinction between the operations (*Handgriff*, grasping by hand) performed by the power of judgment in schematization on the one hand, and concepts (*Begriff*) of the understanding on the other, precisely in order to avoid the aporia of judgment, that is, the infinite need for ever further and higher-order rules to legislate judgment. Consequently, schematization will turn out to be quite literally a grasping by hand, dependent on a hidden art (*eine verborgene Kunst*), with art here referring to a general notion of *techne*, designating the general skill (*Können*) and capacity of grasping by hand. Thus, while first put forward by Kant as a remedy to reason's instrumentality, the canon will be shown to rely on an art, and thus on a technics, in the general sense.

Chapter 2 will reveal the surprisingly central place occupied by instrumentality and technics in Kant's critical project. When it comes to Kant's explicit engagement with the concept, however, technics (*die Technik*) is only mentioned a total of six times throughout the *Critique of Pure Reason*. It is thus all the more remarkable that, in all its grammatical forms, the concept of technics was mentioned a near five hundred times in the *Critique of the Power of Judgment*, published a mere nine years after the first edition of the *Critique of Pure Reason*. Chapter 3 will argue that the *Critique of the Power of Judgment* picks up precisely where Kant had left off in the *Critique of Pure Reason* when he called schematization a hidden art. To the end of securing the relation between concepts and intuitions on the level of the faculties, Kant develops the idea of a technics of nature (*Technik der*

Natur), a Kantian neologism without precedent in the history of philosophy. The idea of technics of nature is to secure and explain how the understanding with its universal laws, can find itself in nature in its actuality and not just in its possibility. But since, properly speaking, we can only judge nature technically in a reflective judgment, and thus with a hypothetical restriction, that which is properly speaking technical will in the end turn out to be our very own power of judgment. Paradoxically, then, the *Critique of the Power of Judgment* harbours the following predicament. On the one hand, Kant employs the term technics in its substantive (*Technik der Natur, Technica speciosa, Technicism*) and adjectival (*technisch*) forms a near five hundred times throughout the *Critique of the Power of Judgment*, indicating that technics most certainly occupies a central and explicit place within the *Critique of the Power of Judgment*. On the other hand, and despite calling nature in its actuality technical hundreds of times throughout the print edition of the *Critique of the Power of Judgment*, for Kant, it is ultimately not nature that is technical but our own power of judgment. But while the unpublished first introduction to the *Critique of the Power of Judgment* explicitly names the power of judgment as the properly technical instance that allows us to judge nature technically, the final print version of the book omits any reference to a technical power of judgment. And as if that was not enough, Kant insists twice that the sphere of purposiveness that is not legislated by the categorical imperative should actually be called technical, seemingly forgetting that he had always already called it so. Thus, what is properly speaking not technical is called technical, while that which is properly speaking technical is not. And what has always already been called technical is insisted upon, again and again, as that which should properly be called technical.

Had Kant's philosophical and authorial task ended with the *Critique of the Power of Judgment*, one might be inclined to agree with Heidegger, Simondon, and Stiegler, even if for different reasons, that Kant's relation to technics was indeed uneasy, as it occupied an important yet difficult role within the *Critique of the Power of Judgment*, oscillating in and out of sight, at once appearing from and then receding into the background. Kant, however, continued to write until shortly before his death in 1804. And it is precisely in the manuscript to his last project, referred to under the name *Opus Postumum* in English language Kant scholarship, that Kant re-evaluates his preceding engagement with and understanding of technics. Chapter 4 will show that the *Opus Postumum* contains Kant's mature systematic reflections regarding the place and role of technics, and specifically technical-practical reason, in the system of transcendental philosophy. The most important change in this regard is that while the idea of a technics of nature was presented by Kant as a necessary regulative supplement for science and thus theoretical philosophy, the *Opus Postumum* will move to place the entire sphere of theoretical philosophy under the jurisdiction of technical-practical reason. Furthermore, technical-practical reason is systematically treated in the last two Fascicles of the *Opus Postumum*, and thus within and following from the context of Kant's renewed engagement with self-positing, that is, the twofold act by which the subject makes itself into an object of experience. While the first act concerns the merely logical act of constituting oneself as an object of thought, I will show that the second, subsequent act of self-positing, by which the subject posits itself as an object in time and space, is ultimately the responsibility of technical-practical reason.

Contra Heidegger, Simondon, and Stiegler, this book thus proposes to articulate the longstanding and systematic engagement with technics at the heart of Kant's philosophical project. Already the *Critique of Pure Reason* must be recognized as orienting itself against reason's technics of instrumentality, as it furthermore reads all philosophy prior to its own critical intervention as the long history of instrumental reason. At the same time, however, and as the discussions of schematization and the discipline of reason will show, the *Critique of Pure Reason* already relied on a positive technics determined in opposition to reason's pathological instrumentality. This other technics will then be further developed, if only tentatively, in the *Critique of the Power of Judgment*, before fully stepping into the light in Kant's final, unfinished piece of writing, the *Opus Postumum*. What this trajectory from the three *Critiques* to the *Opus Postumum* shows is that Kant's transcendental philosophy has always already been *of* technics, at once explicitly constituted against and by it. In the following, then, a surprising image of Kant will step to the fore, as Kant will be shown to have engaged in several simultaneous technical discussions, the importance of which will finally find itself reflected in the outline of the system of transcendental philosophy sketched in the *Opus Postumum*, a system which at once grants technical-practical reason a systematic place while at the same time itself being a first product of technical-practical reason.

Following this excursion on the meaning and role of technics in Kant's philosophy from the *Critique of Pure Reason* until the *Opus Postumum*, the final chapter of this book functions as an *Epilogue* to what has come before, in the sense that the conceptual inquiry into Kant's explicit and manifold engagement with technics has come to an end, but a final question still remains. In a strange turn of events, the manuscript of the *Opus Postumum* provides the scene in which Kant and technics engage in yet another disputation around the question and problem of forgetting. The unfinished manuscript of the *Opus Postumum*, which substantially deals with the question of technics and technical-practical reason, has slowly become the scene of philosophy's long struggle with the threat posed by forgetting; in the end of his life, Kant had become senile. Engaging the question of the relation between philosophy and forgetting by way of Kant's forgetfulness as well as his own thought on forgetting, I will argue that and how, in line with the broader project of the Enlightenment, Kant explicitly favoured forgetting over and above memory, practicing and advocating for a rational technics of forgetting in service of enlightened and autonomous thought and reason.

Part I

Technics, periodization and the eclipse of Kant in the technical thought of Heidegger, Simondon, and Stiegler

The twentieth century was witness to an upsurge of European philosophical thought on technics, addressing both the domains of skills and tools, instruments and machines, as well as technology, the scientific meta-discourse on technics. Among a whole array of perspectives, Martin Heidegger, Gilbert Simondon, and Bernard Stiegler stand out as three of the most lucid and fruitful contributors. A closer look into their projects reveals that each, in their own way, had a multifaceted yet uneasy relation to Kantian philosophy. In one way or another, Heidegger, Simondon, and Stiegler each put forward a symptomatic reading of Kant, according to which Kant articulated the problem of modern technics without self-reflexively understanding the nature of his own insight. Consequently, Kant is given a prominent place in their projects while, at the same time, read as being incapable of recognizing the true nature of technics. The reason for this structural incapacity to think technics on Kant's part is said to be internal to the constitution of transcendental philosophy, leading each of the three thinkers to the conclusion, again and again, that when it comes to addressing the question and problem of technics, there is no future for Kant nor transcendental philosophy.

The aim of this chapter is specific and limited. Before diving into Kant's own texts from Chapter 2 onwards, the present chapter proposes three consecutive introductions of Martin Heidegger, Gilbert Simondon and Bernard Stiegler's technical thought. Each of their accounts will be discussed according to three main points. Each thinker will first be discussed according to their own terms and in respect to their technical thinking,[1]; second via their reading of the history of philosophy and finally through Kant's role within both their thinking and their projected history of technical thought.

1

Heidegger and the question of technics

Martin Heidegger published his seminal essay *The Question Concerning Technics* (*Die Frage nach der Technik*) in 1954. Providing the German pole to the techno-optimistic parallel discourse emerging in France (involving André Leroi-Gourhan, Gilbert Simondon, Jacques Derrida and others), Heidegger's elaborations appear to be more pessimistic in nature. Technics, he writes, 'sets upon man, i.e., challenges him forth, to reveal the real, in the mode of ordering, as a standing reserve. Enframing means that way of revealing which holds sway in the essence of modern [technics] and which is itself nothing [technical].'[1] With these famous words, Heidegger articulated the threat posed by modern technics for *Dasein*. This threat is said to be the consequence of the curiously un-technical essence of technics, with Heidegger here indicating that this will not be a discourse on machines and tools: for Heidegger, if one remains within a so-called 'technical' discourse on technics, primarily concerned with the concrete technical object, 'we remain unfree and chained to [technics], whether we passionately affirm or deny it.'[2] This, Heidegger claims, is due to the relata that are put into relation within the technical discourse on technics – on the one hand, the concrete technical object, and on the other, the human. In order to bring forth a free relation to technics, it is necessary to shift the discussion from the concrete technical object in its relation to the human to that of the essence of technics in its relation to *Dasein*.

While the nature of such a free relation remains yet to be determined, one can nonetheless already begin to qualify Heidegger's supposed techno-pessimism. It appears that Heidegger can neither be said to position himself 'against' technics, not even in its modern guise, nor does this seem to be a project set to 'free' *Dasein* from technics. In his posthumously published notes on technics, entitled *Techne und Technik*, he holds that his 'thinking is not against technics [. . .] but against the superficiality and naivety with which technics is being considered.'[3] In another set of notes, Heidegger further writes:

> To reject technics?
> Is no less foolish than rejecting the sun.
> What is this rejection? – A self-deception![4]

Heidegger's position thus is not directed *against* technics, nor does it aim to free *Dasein* from technics. Instead, his relation to technics seems to be twofold. On the one hand, by mentioning the sun and technics in one stroke, Heidegger proposes technics as a quasi-natural and therefore necessary element of our contemporary world that one would be foolish to try to reject. At the same time, however, and despite the fact that technics is everywhere and in everyone's mouths and minds, it is subject to the general thoughtlessness that Heidegger diagnoses in the historical moment at large. Heidegger's point, then, is that the right question concerning technics has not yet been asked. 'The question concerning technics. The emphasis lies on the question. It is necessary to develop it above all. To endure the question. The question is more pressing than technics itself.'[5] However, in as much as it would be a self-deception to try to reject technics, Heidegger's position should also not be understood as rejecting the current discourses on technics *tout court*. Differentiating between the right and the true, in the sense that the right makes accurate observations without, however, revealing inner truth,[6] Heidegger's aim is to search in and through what is right in order to arrive at the true. Heidegger thus begins his inquiry into the inner truth of technics by working through two common approaches to technics, namely the instrumental approach and the anthropological approach. According to both approaches, technics is a means (an instrument) for the human to bring about an end and thus an essentially anthropological characteristic. And indeed, who could doubt that the human employs technical tools as a means for its ends?

From *techne* to technics

Heidegger begins his discussion of the anthropological and instrumental discourses on technics in reference to Aristotle, who had famously determined the ancient Greek notion of *techne* in a twofold manner. On the one hand, Aristotle approached *techne* in relation to nature (*physis*) and thus within an ontological discussion of technical bringing forth (*poiesis*). On the other hand, *techne* was also discussed in contradistinction to the systematic and unchanging Greek notion of science (*episteme*), its epistemological counterpart. Heidegger's reading of the Aristotelian text proposes that what is at stake in both of these relations is in fact revealing (*entbergen*, *aletheia*), meaning that in both relations *techne* is a specific mode of revealing. Technical bringing forth (*poiesis*) reveals things which cannot come into appearance by themselves, in as much as any technical form of knowledge is nothing but a way of opening up and thus a mode of revealing once again.[7] Starting from the instrumental and anthropological definitions of technics as instruments and means for human ends, Heidegger thus arrives at revealing as the inner truth of *techne*. But can revealing be said to remain the key to understanding technics in its modern guise as well?

In the posthumously published collection of notes entitled *Techne und Technik*,[8] Heidegger explicitly deals with the relation between the ancient *techne* and modern *Technik*. Here, Heidegger writes that the machine (*die Kraftmaschine*) is precisely not the '[. . .] imitation of handicraft and the natural process'.[9] The machine, he writes, has a 'motoric character which is essentially related to the generation of power',[10] meaning that it generates *Kraft* and thus energy rather than force, since it is thermodynamic in character. While the above section from *Techne und Technik* was already drafted around the year 1940, the same argument will be employed in the later *The Question Concerning Technics*. Modern technics is here primarily characterized as a demand directed at nature 'to supply energy that can be extracted and stored as such'.[11] In its essential demand of generating and storing energy, modern technics is thus fundamentally distinguished from its pre-modern instantiation.

At the same time, however, Heidegger claims that modern technics remains a mode of revealing (*entbergen*) nonetheless, meaning that it nevertheless retains a direct relation to the ancient form of *techne*. But rather than revealing as a mode of bringing forth (*poiesis*) or technical knowledge, modern technics reveals as a mode of challenging (*herausfordern*).[12] Modern technics, however, challenges the real to reveal itself as something specific, which Heidegger calls standing reserve (*Bestand*). What this means is that when challenged to reveal itself by modern technics, the real no longer comes into view as an object (*Gegenstand*) but as a standing reserve (*Bestand*). The modern world, Heidegger claims, can only come into view by being put in reserve, subjected to acts of 'unlocking, transforming, storing, distributing and switching'.[13] As a standing reserve, the real no longer stands against us the same way that an object (*Gegenstand*) does, but instead has been 'ordered to stand by, to be immediately at hand, indeed to stand there just so that it may be on call for a further ordering'.[14] Heidegger's point, then, is that thinking an airplane, for example, with the concept of object (*Gegenstand*) is certainly right. But it fails to be true in the sense that one misrecognizes both what and how the airplane 'is', in the sense of exists. The name standing reserve (*Bestand*) thus 'designates nothing less than the way in which everything presences that is wrought upon by the challenging revealing'.[15] The mode of existence of that which has been challenged and revealed by modern technics is subsequently no longer objective, but a standing reserve (*Bestand*).

When it comes to the difference between ancient and modern technics, there is a second important difference that Heidegger points out. Complementing the relation between, or rather the move from, object (*Gegenstand*) to standing reserve (*Bestand*), Heidegger situates the conceptual relation between mind (*Gemüt*) and enframing (*Gestell*). The *Gemüt*, Heidegger writes, is 'that original gathering from which unfold the ways in which we have feelings of one kind or another'.[16] Enframing (*Gestell*) is equally said to have a gathering or assembling function. But rather than assembling our faculties in relation to a world of objects (*Gegenstände*), enframing (*Gestell*) challenges the human to reveal the real as a standing reserve (*Bestand*). And, finally then, it is this call to enframing, according to Heidegger, that is the curiously un-technical essence of modern technics: 'Enframing means that way of revealing which holds sway in the essence of modern [technics] and which is itself nothing [technical]'.[17] It is nothing technical because, in the way that the essence of a tree is not simply another tree and thus cannot be found in a forest, the essence of modern technics is not just another technical object neither.[18] Enframing (*Gestell*), as the essence of technics, thus cannot be found among technical objects, nor adequately addressed from within a technical discourse on technics.

The second term: The human

The preceding section has provided a first, initial discussion of Heidegger's thinking of technics. But in order to round off the preceding discussion and before moving on to Heidegger's account of the history of technical thought, technics needs to be positioned in relation to the human as well. Since enframing (*Gestell*) sets upon or challenges the human to reveal the real as a standing reserve (*Bestand*), the human as the second relata of the technical relation has been called into question. Heidegger's posthumously published collection of notes on technics includes the following remarkable passage on this point. 'The essence of technics is nothing human, but technics belongs to the human because the human belongs to [*in*] the essence of technics. Depending on the essential form [*Wesensgestalt*] of technics, the affiliation of the human to [*in*] the essence of technics is different and, accordingly, the nature of his technical doing is different.'[19] While the essence of technics is here explicitly said to be neither technical nor human, the human is at the same time said to 'belong' to or rather in the essence of technics in different, that is, changing, ways. Heidegger's point here is that there is no immediate (unmediated) relation between 'the human' and technics.[20] Depending on the specific (historical) form of the essence of technics, the human is called upon differently and consequently partakes in the essence of technics in different ways.

This very point was already made by Ernst Jünger in his 1932 essay *The Worker* (*Der Arbeiter*), where he claimed that 'man is bound up with [technics] not directly, but indirectly'.[21] For Jünger, the relation between the human and modern technics is mediated by a metaphysical *Gestalt* (figure), which he named the worker. Technics is then the mode in which this *Gestalt* of the worker mobilizes the world. Heidegger engaged with Jünger's writing throughout his life and was particularly fond of *The Worker*. Band 90 of the *Heidegger Gesamtausgabe* contains the posthumously published collection of Heidegger's notes on Jünger, where we find an important passage from 1954, the same year that *The Question Concerning Technics* was published. Here, Jünger's claim about the mediated relation between the human and technics is explicitly discussed in contradistinction to a conference paper and essay written by the quantum physicist Werner Heisenberg, which proposes that 'the proposition that the human only faces itself is valid in a much broader sense in the age of technics'.[22]

Responding to Heisenberg, Heidegger asks 'who is this man – that meets himself?' A few lines further down, he continues by asking, '*man*, whether such a thing is possible at all? And if thought possible – if not in this opinion the greatest delusion.'[23] *On the Question Concerning Technics* explicitly references the same sentence from Heisenberg, but Heidegger's own answer omits Jünger's name. Heidegger's claim is that when called upon by enframing (*Gestell*) to reveal the real as a standing reserve, the human is itself threatened with disappearing into the standing reserve. And once faced with this threat, 'man, precisely as the one so threatened, exalts himself to the posture of lord of the earth. In this way the impression comes to prevail that everything man encounters exists only insofar as it is his construct. This illusion gives rise in turn to one final delusion: It seems as though man everywhere and always encounters only himself.'[24]

Heidegger's point here is that it only seems as if modern technics has made the world into the mirror image of the human. But in fact, the human is called upon by enframing (*Gestell*) in a specific way, which Jünger called the figure of the worker and which Heidegger can here be seen to call the figure of the lord of this earth (*Gestalt des Herrn der Erde*). Being mediated in its relation to the world by this figure means that the human can no longer reflect itself anywhere in its inner being, meaning that it is essentially endangered by modern technics in its truthful relation both to itself as *Dasein* as well as to everything else that exists.[25] Everywhere we look, we find ourselves reflected back in the figure of the worker as the lord of this earth. This is the case, Heidegger explains, because, while modern technics is indeed a mode of revealing, it is also a mode of veiling and thus of covering. And as such, it covers not only the past mode of revealing as bringing forth (*poiesis*), from which it itself originated, but it furthermore covers revealing as such, 'wherein unconcealment, i.e., truth, comes to pass'.[26] This predicament, which Heidegger also calls *Seinsvergessenheit* and, in its aggravated form *Seinsverlassenheit*, is diagnosed by Heidegger as the central problem of metaphysics ever since ancient Greece.[27]

Heidegger, Jünger, the nature of the historical present and what is to be done

Band 90 of the *Heidegger Gesamtausgabe* shows that Heidegger engaged with Ernst Jünger's writing from as early as 1932, an engagement which Heidegger continued throughout his life.[28] With both thinkers living through the far-reaching technological change imparted on the world by the industrialized slaughter of the First and Second World Wars, including nuclear warfare, and the emergence of Fordism and the production line, Heidegger saw Jünger as someone who experienced these large-scale technological developments on the front line while also having the ability to articulate their inner truth. For Heidegger, Jünger experienced his life as an active soldier during the abysmal First World War essentially through the fundamental insight that was first carved out by Nietzsche's metaphysics.[29] Jünger, he writes, 'gained a foothold in the reality that Nietzsche has suffered in thought'.[30] At the same time however, Heidegger does not see in Jünger 'only' a soldier who lived and suffered the material reality of what Nietzsche had anticipated in thought, but also someone who then, in a second step, was able to find 'the strength and determination of contemplation [*Besinnung*] and the word'.[31] Jünger is thus considered at once as a highly decorated soldier during the *Materialschlacht* (battle of matériel) that was the First World War and a literary genius who was able to give word to 'the historical ground of the essence of the completed age of modernity'.[32] Jünger's writing, and in particular his essay *The Worker*, is read by Heidegger as the key to reflecting on the historical present in its innermost truth. But how is one to understand Heidegger's claim that Jünger articulated the *completed* epoch of modernity when Jünger himself explicitly aimed to bring a *new* reality into view, which he believed to have just begun and which he discussed under the name of the worker, total mobilization and will to power?

Despite the explicit aim and content of Jünger's *The Worker*, which tries to articulate the beginning of a new kind of power (the worker), as well as an epoch (determined by work), which was to destroy bourgeois society, Heidegger claims that Jünger did not articulate the beginning of anything new but that his contribution was rather the astute and lucid articulation of the completion of the modern epoch.[33] For Heidegger, his time was not at all the beginning of anything new, least of all a new epoch. Instead, he understood his time to be the fulfilment of a process that had long been underway and that should now, finally, come to a close under the name of modern technics. Thus, while for Jünger, his time was the beginning of a new millennium and epoch, Heidegger believes Jünger, including *The Worker*, to have articulated the conclusion and end of the modern epoch. For Heidegger, Jünger greatly misinterpreted the temporal nature of his insight. At the heart of Heidegger's reading of Jünger thus lies an underlying disagreement with Jünger's interpretation of the historical nature of their time.

Further to this point stand the two authors' respective methodologies and the resulting self-reflexive conclusions regarding their own function and vocation within the historical present. There are, Heidegger claims, two possible relations to Nietzsche. One can either take root and settle in the reality (*Wirklichkeit*) as will to power that was first revealed to us by Nietzsche. Or one can try and overcome this reality.[34] According to Heidegger, Jünger's contribution was precisely to take root within the space first carved out by Nietzsche.[35] From within that space, Jünger begins to describe what he sees and experiences, and this he is said to do in the clearest and most precise of ways. That same astute use of observation and description, which is admired on the one hand, is however on the other hand also the focus of Heidegger's point of critique. Description, Heidegger argues, 'inscribes ourselves into what belongs to this real [*diesem Wirklichem*]'.[36] The problem that ensues for Heidegger is that Jünger himself takes on the character of that very reality which he is trying to describe. Subsequently, his thinking is said to become calculation (*Rechnen*),[37] until finally, Heidegger argues that Jünger's thinking itself gains the 'character of work [Arbeitscharacter]'.[38] Heidegger's diagnosis of Jünger's thought as work, calculation and, in the last instance, technical/instrumental reason leads him to formulate the following conclusion. While on the one hand, Heidegger charges Jünger with first giving word to 'the historical ground of the essence of the completed era of the modern age',[39] he is at the same time critiqued and characterized as suffering from the essence of modern technics, that is, enframing, whereby Jünger's work itself must be read as a symptom, rather than a self-reflexive account of modern technics. Jünger is said to 'carry out [. . .] the lead [*Hinführung*] to the real [*das Wirkliche*] of the will to power and is ruthlessly serious about this real'.[40] Moreover, Heidegger charges Jünger with being guilty of having brought about 'the utmost culmination and unrestrained expansion and unleashing of what we have to recognize as the reality of the "modern age" [*Neuzeit*], which has already lasted for 300 years'.[41] In so perfectly and precisely describing the reality he finds himself in, Heidegger argues that Jünger ends up affirming, furthering, and unleashing the very reality he initially set out to critique.

What becomes clear at this point, then, is that Heidegger's ruthless critique of Jünger allows for insight into Heidegger's account of the historical present as well as his own vocation within that present. The question of what can and needs to be done as well as in what way is pressing for Heidegger, who understands his time to be that of a transition (*Übergang*). On one side of this transition, he places metaphysics, which is completing itself. On the other side of the transition, he places a 'new time', which, however, has not yet begun.[42] Our vocation, he then writes, is to bear witness to this transition. Heidegger's characterization of his time as transitory in nature illuminates why *The Question Concerning Technics* self-reflexively understood itself as merely preparing (*vorbereiten*) a new, and ultimately free, relation between technics in its essence and *Dasein*. In contradistinction to Jünger, Heidegger's methodology aims to prepare rather than describe a different time through the act of questioning. 'If we are able to inquire into the essence of technology properly, the questioning changes into the discussing saying [*das erörternde Sagen*], the thinking changes. We no longer move in the representation of objects [. . .]. We are on the way – where to? Such thinking is no longer enframeable [*stellbar*] in and through enframing [. . .].'[43] Heidegger's proposition is thus, that asking about the essence of technics sends us on a journey towards a different kind of thinking, which, different to Jünger's methodology, is precisely not enframeable (*stellbar*) and whose endpoint cannot be predetermined from where we currently find ourselves. What is clear, then, is that the path that Heidegger aims to embark on could not stand in starker relief to Jünger's descriptive, laborious method, which is said to have not only greatly misinterpreted the temporal nature of the historical present, but further, in this misinterpretation, ended up affirming and unleashing the very reality he set out to critique.[44]

Heidegger's history of technical thought: From metaphysics to technics

Despite Heidegger's ruthless critique of Jünger's method and self-reflexive position within the movement of history, Heidegger nevertheless grants Jünger an important position within his account of the history of technical thought. In the previous sections, *The Question Concerning Technics* was seen to differentiate pre-modern *techne* from modern technics (*Technik*) by a twofold move from object to standing reserve (*Bestand*) and from mind (*Gemüt*) to enframing (*Gestell*). We thus already know something about the specificity of Heidegger's conception of modern technics vis à vis its pre-modern instantiation. At no point, however, have we determined how exactly this transition from the pre-modern to the modern supposedly occurred. In order to engage with Heidegger's periodization claims, it is important to keep in mind that he differentiates between the chronologically right and the historially (*geschichtlich*) true.[45] Thus, when considered chronologically, the emergence of modern technics post-dates the emergence of modern natural science because modern technics is dependent on the exact natural sciences for its functioning.[46] Subsequently, modern technics emerged two centuries after the advent of modern science with the invention of the modern machine (*Kraftmaschine*) around the year 1750. This chronological history of how the emergence of modern science is followed by modern technics is most certainly right. There is plenty of material evidence in the form of concrete technical objects to support it. But is it also true?

Heidegger's critique of the chronological periodization of the advent of modern technics hinges, once again, on the difference and overlap between the concrete technical object and the essence of technics. While the modern machine, meaning the concrete modern technical object, most certainly matters, it is necessary to keep in mind that it is the phenomenal manifestation of the essence of modern technics. And this very essence, Heidegger claims, already 'holds sway' in physics. 'Modern physics is the herald of Enframing, a herald whose origin is still unknown.'[47] The essence of modern technics is thus, according to Heidegger, already at work in the modern exact sciences, which emerged during the seventeenth century, before coming into the open with the modern machine. What this means, then, is that what chronologically appears as later (i.e. modern technics), in truth, when considered historially and in terms of its inner essence, predates modern science.

> Because the essence of modern technology lies in Enframing, modern technology must employ exact physical science. Through its so doing, the deceptive illusion arises that modern technology is applied physical science. This illusion can maintain itself only so long as neither the essential origin of modern science nor indeed the essence of modern technology is adequately found out through questioning.[48]

Heidegger thus claims that the essence of modern technics, enframing, first emerged during the seventeenth century with modern physics, after which, two centuries later, the concrete technical object emerged, by way of modern natural science, in the form of the modern machine.

In his posthumously published notes on technics, Heidegger further qualifies the relation between modern technics and modern science. Here, they are both said to derive from the same metaphysical root (*dieselbe metaphysische Wurzel*).[49] Consequently, discussions concerning the priority or dependence of technics on science are futile and nonsensical, since they both derive from metaphysics.[50] At the same time, there is an asymmetry between metaphysics' two descendants, since modern technics does not only derive from metaphysics but is furthermore 'the actual completion of <metaphysics>'.[51] In its last stage, metaphysics is thus said to determine itself as technics.[52] Subsequently, and when considered from the historial rather than the chronological point of view, the question concerning technics turns out to be nothing but the question concerning metaphysics, as technics here becomes the question concerning the 'truth in which beings are positioned (this positioning as the destiny of Being [*Seynsgeschick*] itself) and according to the Being [Seyn] that has abandoned beings into such truth'.[53] What this means for our purpose here, then, is that our inquiry into Heidegger's account of the history of technical thought demands an inquiry into the history of metaphysics, as it is in its final stage that metaphysics 'becomes' technics.[54]

However, working with a category as broad as metaphysics goes beyond the limits of this present discussion. The following discussion will subsequently employ a more restricted notion of metaphysics, namely that of modern metaphysics and with it, philosophical modernity. In his notes on Jünger, Heidegger writes that philosophical modernity was 'opened' with Machiavelli's 1532 *Il Principe*. The closing point, he goes on to argue, is marked by none other than Jünger's *The Worker*.[55] In between Machiavelli and Jünger, that is, in between the prince and the worker, Heidegger positions Descartes and Newton. Descartes is said to have performed the 'foundation of modern mankind'[56] and modern thought in 1637 with his *Discourse on the Method*, the essential feature (*Grundzug*) of which, according to Heidegger, is 'the mathematical' (*mathemata*).[57] While the first effects of the mathematical are argued to have arisen during the late scholasticism of the fifteenth century, and thus sufficiently earlier than the previously mentioned seventeenth century, Descartes contributions in the seventeenth century allowed for the mathematical to be further clarified and determined and Newton is credited with having subsequently produced the 'first systematic and creative conclusion'[58] with his *Principia Mathematica* in 1686–7.

In the next step, Heidegger then goes on to introduce a new protagonist into the question of metaphysics qua the question of technics. Immanuel Kant, and specifically his 1781/1787 *Critique of Pure Reason*, is introduced by Heidegger as performing a crucial role within the movement of modern metaphysics to technics and exemplifies Heidegger's broader twofold relation to the history of technical thought. Heidegger claims that from its outset, the *Critique of Pure Reason* already takes place in a world of mathematical-physical objects, never even questioning whether there could be another access to the world of things than the one prescribed by Newtonian science. The *Critique of Pure Reason* is thus said to essentially lay open and further inscribe the modern foundational attitude (*Grundstellung*), which furthermore remains the 'basic historical and spiritual stance [*geschichtlich-geistige Grundstellung*], which supports and determines us today'.[59]

Heidegger and Kant

The important question here is, then, how to conceive of the relation between the mathematical, as the characteristic feature of modern metaphysics, and what was discussed earlier as enframing, that is the essence of modern technics. The work best suited to addressing the question of the relation between enframing and the mathematical in the Heideggerian corpus is without doubt his 1962 *The Question Concerning the Thing*, the reworked, published transcript of what was initially presented by Heidegger as a lecture course in the winter semester of 1935–6. This work is thus situated at once before and after the 1954 *The Question Concerning Technics*, to which it stands in a quasi-complementary relation. *The Question Concerning the Thing* not only asks the metaphysical question *par excellence* – what is a thing? – but narrows this question: it asks what is and how something can become a thing for us in the modern period. And since, as shown earlier, the standing reserve is seen as the mode of existence of things subjected to enframing, the question is, then, if and how *The Question Concerning the Thing* allows us to narrate the history of metaphysics as the movement from the objective mode of existence (*Gegen-Stand*) to that of the standing reserve (*Be-Stand*).

Heidegger begins his inquiry into the thing by pointing out the intrinsic historial dimension of thingness. The old metaphysical question 'What is a thing?' here immediately turns out to be a historial question, asking how something can become a thing for us in and for a specific time. The hypothesis put forward by Heidegger is that when it comes to the modern age, the 'thingness of things' is determined by the mathematical (*mathemata*), which is 'that "in" things [*jenes >>an<< den Dingen*] which we really already know; hence, what we do not first have to fetch from things, but what we bring along with us in a certain way'.[60] The mathematical is thus the thingness of things, which *a priori* allows things to come into view as things for us in the modern period.

From the point of view of the mathematical, 'the givens of everyday getting around in the world [*das umgänglich alltägliche gegebene*] are construed as mere material and splintered into a manifold of sensations',[61] which, once ordered and organized, can then come into view as an object of mathematical-physical science. In *The Question Concerning the Thing*, 'material' thus designates material for the *a priori* forms of the mathematical sketch of the thingness of things. But, is this consideration of something as sheer material not also precisely what Heidegger had in mind in *The Question Concerning Technics*, when he wrote that 'man's ordering attitude and behavior display themselves first in the rise of modern physics as an exact science?'[62] The notion of 'material' is here the hinge between both the mathematical sketch of the thingness of things, which orders 'material' according to its a priori forms, and enframing, which challenges the real to reveal itself as a standing reserve and thus as 'material' for further ordering and organization. Reading Heidegger's *The Question Concerning the Thing* with and through *The Question Concerning Technics*, we can see that Kant's *Critique of Pure Reason* describes nothing less than the ordering attitude (*bestellendes Verhalten*) of the mathematical sketch for Heidegger, which in *The Question Concerning Technics* is also called enframing.

Heidegger charges Kant with numerous further contributions to the movement of modern thought towards modern technics. First, and most importantly, as we have already seen, Kant is read as having formulated the 'basic historical and spiritual stance, which supports and determines us today'.[63] This basic historical and spiritual stance (*Grundstellung*) is essentially modern and determined by the mathematical. The problem with a *Grundstellung*, however, is that it needs to be investigated and inquired into in order to come out into the open and be understood.

> In mastering or failing to master the question of the thing in a sufficiently thoughtful way, or in disregarding it altogether, there are decisions whose field of play and distance in our history are to be considered always only after centuries [have passed]. The confrontation with Kant's step should provide us with the proper perspective [*das rechte Augenmaß*] for such decisions.[64]

Second, then, Kant is not only said to have articulated the modern *Grundstellung*, he is furthermore said to provide the proper perspective (*Augenmaß*), and thus methodological instrument, to account for how the decisions taken within that modern foundation stance have shaped and continue to shape the present. Kant is thus at once the subject and means of Heidegger's inquiry into modern thinking.

Third, Kant's *Critique of Pure Reason* is read by Heidegger as formulating the ordering attitude (*bestellendes Verhalten*) of the mathematical sketch, which in *The Question Concerning Technics* is called enframing (*Gestell*).

There is, however, a fourth important point to Heidegger's discussion of Kant. This last point concerns nothing less than the movement from metaphysics to modern technics itself. With the *Critique of Pure Reason* mapping onto the system of general and special metaphysics of the Wolff-Leibniz school, the chapter entitled *Transcendental Analytic* deals with what was hitherto discussed under the name of general metaphysics, and thus ontology: Kant's self-professed aim with the *Transcendental Analytic* was to replace 'the proud name of ontology [with] the more modest title of a transcendental analytic'.[65] At the same time, the *Transcendental Analytic*, and specifically the section entitled *Analytic of Principles*, is the very place in which Kant lays out the mathematical sketch of the thingness of things. When considered from the perspective of Heidegger's reading of Kant performed in *The Question Concerning the Thing*, the Kantian move from metaphysics to the *Transcendental Analytic* is thus, in essence, the very moment in which modern thought self-reflexively did away, so to speak, with metaphysics. This moment as such shifted the discussion of metaphysics to the mathematical sketch of the thingness of things, which at heart is nothing but enframing (*Gestell*) and thus the essence of modern technics. Kant's *Transcendental Analytic* is thus, in a sense, the moment in which metaphysics 'became' technics. Consequently, thinking the Kantian intervention from Heidegger's point of view, the *Transcendental Analytic* could or rather should have been called *Transcendental Technics*.

Reflecting on our trajectory so far, we can see that and how Heidegger reads the history of technical thought as a twofold movement from the ancient Greek *techne* to modern technics. Modern technics is further discussed and problematized in relation to modern metaphysics. Within the movement of modern metaphysics to technics, Kant situates Machiavelli and Jünger, understood as the respective opening and closing figures of modern metaphysics. Within that movement, Heidegger also assigns important places to Descartes, Newton and finally Kant. Kant's role is explicitly discussed in *The Question Concerning the Thing*, which was considered in its complementary relation to *The Question Concerning Technics*, as both texts, in their own way, ask the metaphysical question *par excellence*: What is a thing? Kant clearly occupied a multifaceted role. Apart from providing insight into the modern historical and spiritual stance (*Grundstellung*), he was also discussed as offering the proper perspective (*Augenmaß*) for adequately understanding this modern stance. Third, *The Question Concerning the Thing* laid out the ordering attitude of the human, which had first showed itself in the natural sciences. Here, the notion of 'material' was discussed as the hinge between both the mathematical outline, which orders 'material' according to its a priori forms, and enframing, which challenges the real to reveal itself as a standing reserve and thus as 'material' for further ordering and organization. Fourth and most importantly, I argued that the Kantian move to do away with metaphysics in favour of the *Transcendental Analytic*, which laid open the a priori principles of the thingness of things, provides the moment in which metaphysics, so to speak, 'became' modern technics within the Heideggerian schema. Kant's *Transcendental Analytic*, in Heidegger's reading, is thus in essence a *Transcendental Technics*.

On the one hand, then, we can say that from the perspective of Heideggerian thought, Kant marks the moment in which metaphysics 'became' technics. On the other hand, the fact that Kant called his intervention *Transcendental Analytic* rather than *Transcendental Technics* means that for Heidegger, Kant ultimately misunderstood the nature of his own intervention. Must Kant, in a similar way to Jünger, who in the Heideggerian account finally concluded and fulfilled the move from modern metaphysics to technics, then not also be read as having described and further inscribed the modern foundational attitude? Kant's explicit aim in the *Critique of Pure Reason* is to show that and how there can be no other access to the world of objects than the one prescribed by the conditions of possibility of the objects of experience put forth in the *Transcendental Analytic* and which Heidegger calls the mathematical sketch of the thingness of things.[66,67] This is the case because, according to the Kantian so-called 'Copernican revolution', it is the objects that must conform to our forms of cognition, rather than the other way around. Kant thus essentially 'suffers' from enframing, rather than thinking it, as he inscribes and consolidates the modern foundational attitude as our only possible access to the world of things. Consequently, the insight into the mathematical and enframing provided by Kant is merely symptomatic, transforming Kant into a symptom, possibly *the* symptom of modern technics, who articulates, but only implicitly, the 'basic historical and spiritual stance [*geschichtlich-geistige Grundstellung*], which supports and determines us today'.[68]

Consequently, then, Kant occupies a central but twofold role in Heidegger's account of the movement of metaphysics to technics. While Kant articulates the modern foundational stance, he is at the same time not read as a thinker of technics. While providing the foundation for Heidegger's own technics-thinking as enframing (*Gestell*), Kant is at the same time credited with no 'proper' insight into it. Essentially, however, Heidegger's symptomatic reading of Kant only makes sense under the assumption that Kant never put forth any explicit writings on technics himself, or if there were writings on technics, that Kant did not recognize their inherent technical nature. As the following chapters will show, Kant did, however, 'think' technics, despite Heidegger's refusal to acknowledge it. In fact, the German philosopher and editor of the *Opus Postumum*, Gerhard Lehmann, pointed out as early as 1938 in his habilitation that Kant's last, unfinished manuscript, the *Opus Postumum*, contained the manuscript of what would have become Kant's critique of technical-practical reason, had he lived to complete it.[69] The fact that Heidegger referred to the *Opus Postumum* on numerous occasions[70] proposes that this eclipse was not simply a matter of Heidegger 'not doing his reading'. Furthermore, any quick defence of Heidegger's omission by way of the argument that Kant's explicit technical thought might simply have contained a 'bad' technics – further proving that Kant indeed forgot the question of Being as technics – has, since Lehmann, also been put into question by Yuk Hui. Hui follows Gilbert Simondon in arguing that it was in fact already in the earlier 1790 *Critique of the Power of Judgment* that Kant explicitly 'thought' technics.[71] Rather, the preceding engagement with Heidegger aims to suggest that it was in fact Heidegger's damning diagnosis, according to which the right question of technics had not yet been asked, and his consequent self-positioning within the history of philosophy, which led Heidegger, in a somewhat tragic turn of events, to further conceal technical thinking prior to his own intervention. For Heidegger, the history of philosophy, lucidly demonstrated by his reading of Kant, is read as entirely devoid of explicit technical thought proper, while it implicitly harbours the seeds for Heidegger's technical thought alone. What remains to be done, then, is to inquire into Kant's thinking of technics, which might just reveal, in the process of doing so, one thing or another about Heidegger's own technics-thinking and where it might have come from.

2

Simondon and the problem of technics

Martin Heidegger was not the only philosopher to put forward an inquiry into the essence of technics in the post-war period. Gilbert Simondon's *On the Mode of Existence of Technical Objects* (1958)[1] can be read as the French techno-optimistic counterpart to Martin Heidegger's intervention in *The Question Concerning Technics*.[2] In contrast to Heidegger's explicitly non-technical engagement with the question of technics, the explicit assumption of Simondon's title already suggests that according to Simondon, there is precisely such a thing as a technical mode of existence, which he calls *réalité technique* (technical reality).

In order to fully appreciate the scale and depth of Simondon's intervention in and engagement with the Heideggerian proposition, an important translation issue needs to be taken into consideration. In the first 1939 French translation of Heidegger's *Qu'est-ce que la métaphysique?* (*Was ist Metaphysik?*), Henry Corbin translated Heidegger's notion of *Dasein* as *réalité humaine*.[3] Following and at the same time diverting from the Heideggerian path, Simondon ventured to employ the term *réalité humaine* and furthermore supplement it with his own invention: *réalité technique*. Cecile Malaspina and Jon Rogove, the translators of the 2017 Univocal English translation of *On the Mode of Existence of Technical Objects*, have chosen to translate the two terms directly, as 'human reality' and 'technical reality'. However, rather than referring to reality as a philosophical category of quality, Simondon must be seen to fundamentally refer to human and technical being (*Dasein*). Employing the Heideggerian terminology in its specific French philosophical history, Simondon argues that technical objects have being (*Dasein*) – a properly technical mode of existence that he calls technical reality – which is further complemented by a properly technical essence (technicity). At this point, it should be clear that we are indeed a long way from Heidegger's famous claim, in *The Question Concerning Technics*, that 'the essence of [technics] is by no means anything technical'.[4] Simondon's thesis is that there is, essentially, such a thing as technical being, and that technicity only provisionally and partly manifests itself in the material technical objects that we interact with in the world. 'Objects appear at a certain moment, but technicity precedes them and goes beyond them; technical objects result from an objectivation of technicity.'[5] Technicity must thus be thought of as a power or potentiality in the proper sense, 'a depository of a capacity to evolve'.[6]

Simondon's accounts of technical reality and technical essence are not, however, the only fundamental differences between the Simondonian and the Heideggerian discourse. The previous section on Heidegger's technical thought showed that a discourse of questioning runs through Heidegger's philosophy. Simondon's discourse, on the other hand, centres not so much on questions but instead on the problem and the problematic. However, the problematic is here not primarily an epistemological trope formulated by a thinking subject, as Gilles Deleuze pointed out in his 1966 review of Simondon's *L'Individu et sa genèse physico-biologique* (Individuation and its Physical-Biological Genesis).[7] Rather, the problem 'acquires in Simondon's thought tremendous importance in so far as the category is endowed with an objective sense'.[8] Simondon thus extends the territory of the problem to include objective reality, as he understands the problem to be the main dynamism in both the becoming of the world and the main figure through which an understanding of this becoming of the world comes about. What this means, then, is that becoming, or genesis, is at heart the resolution of a problem. Becoming, Simondon writes in the Introduction to *Individuation in Light of Notions of Form and Information*, 'is not a framework in which the being exists; it is the being's dimension, the mode of resolution of an incompatibility that is rich in potentials'.[9]

Consequently, Simondon also takes issue with the metaphysical discipline of ontology. But differently to Kant, who replaced 'the proud name of ontology [with] the more modest title of a transcendental analytic',[10] and differently to Heidegger, who derived his notion of enframing (*Gestell*) from reading Kant's *Transcendental Analytic* in favour of his own *Existential Analytic*, Simondon took another route. Simondon replaced the metaphysical discipline of ontology with that of ontogenesis, understood as an inquiry into ontogenetic problems, and resolutions, understood as that 'through which the being becomes insofar as it is, qua being'.[11] What, then, is the problem of technics as formulated by Simondon in *On the Mode of Existence of Technical Objects*, and how does technical being become?

The problem of technics and becoming in Simondon's *On the Mode of Existence of Technical Objects*

Philosophy, Simondon claims, has historically approached the reality of being in two, mutually exclusive, ways. First, there is the substantialist mode of approach, which holds that being is not subject to construction or creation but is rather given to itself or founded on itself in its unity.[12] Second, there is the hylomorphic mode of approach. Hylomorphism is the ancient philosophical doctrine of how form (*morphe*), usually taken to be active, bestows its shape on passive matter (*hyle*). According to hylomorphism, the individual 'is generated by the encounter of a form and a matter'.[13] What both of these approaches share is that they 'suppose that there is a principle of individuation prior to individuation itself, that is capable of explaining, producing, and guiding it'.[14] What this means is that both philosophical doctrines take a given individual as their starting point in order to then work their way back to its condition of existence, whereby granting ontological privilege to the constituted individual. Simondon's project aims to critique both approaches as he instead ventures to understand the individual through individuation rather than the other way around.[15]

Within the more restricted context of Simondon's technical thought, one encounters the same opposition and problematization of philosophical doctrines in the third and final chapter of *On the Mode of Existence of Technical Objects*. Here, the substantialist first pole is termed *Apriorism*, which is said to attempt to grasp technics by means of the idea and proceeds according to the logical procedure of deduction. The primary representative of *Apriorism* mentioned by Simondon is Plato.[16] The hylomorphic approach supplies the corresponding pole of *Aposteriorism*, which employs concepts over ideas and is generally said to be the complementary empiricist, conceptual and partly nominalist approach to technics (and philosophy more broadly).[17] Discussing both approaches, Simondon employs a temporal critique, arguing that metaphysical thought in both instantiations is either too early (a priori) or too late (a posteriori), but never contemporary with the becoming of technics. The temporal problematization of these two philosophical approaches can already be found in *Individuation in Light of Notions of Form and Information*, where he writes that 'the search for the principle of individuation is finished either after individuation or before individuation, depending on whether the model of the individual is physical (for substantialistic atomism) or technological (for the hylomorphic scheme)'.[18] Despite Simondon's claim that metaphysics is structurally incapable of accounting for the mode of existence of technical objects due to its being out of time with the becoming of technics, his argument at the same time suggests that hylomorphism is essentially based on a technical model of individuation. How is one to understand this argument, where metaphysics is at once incapable of thinking technics and yet technical at heart?

Simondon's point concerning hylomorphism is that its supposed universal and logical scheme is in fact nothing but 'the transportation into philosophical thought of the technical operation reduced to work'.[19] The main critique of hylomorphism, then, is that it leaves the active centre of the technical operation obscure and instead relies on the activity of human work to effectuate the link between the two terms of matter and form. But while, as in brickmaking, for instance, the worker surely prepares the clay in order for it to take on form, it is 'the clay that takes form according to the mold, not the worker who gives it its form'.[20] Simondon's point is thus that at the foundation of the hylomorphic scheme, and thus at the foundation of the *Aposteriorism* of operational knowledge, lies a specific but ultimately incomplete technical experience.[21] As he puts this elsewhere,

> the technical operation that *imposes a form on a passive and undetermined matter* [. . .] is essentially the operation controlled by the free man and executed by the slave; the free man chooses the matter – which is undetermined because it suffices to designate it generically by the name substance – without seeing it, without manipulating it, and without preparing it.[22]

Consequently, hylomorphism considers form to be the active principle of individuation, as can be seen at work from Aristotle's *Physics* to Hegel's lord in the *Phenomenology of Spirit*, because 'the veritable passivity of the matter is its abstract availability behind the given order that other men will execute'.[23] It is this simplified and socio-politically overdetermined model of technical production that is ultimately the target of Simondon's critique. While attention is paid to the two terms of matter and form, or passivity and activity, hylomorphism leaves what happens between them in the dark. Looking at the centre of this operation would reveal, however, that it is in fact the clay that actively keeps the form embodied in the mould of the brick, rather than the worker transferring form onto passive matter in the name of the master.[24] While for Heidegger modern technics was the completion of metaphysics, it is now clear that for Simondon an incomplete technical experience, overdetermined by the socio-economic and political reality of the ancient Greek master-slave dynamic, lies at the heart of the so-called abstract logical categories of the hylomorphic metaphysician ever since antiquity.

Simondon's proposition, then, is to replace this inaccurate and incomplete techno-metaphysical schema of hylomorphism with a detailed account of technical operation proper. This move is both conceptual and historical in nature. Simondon argues that the metaphysical schema of the universalized master's perspective could only remain unquestioned for as long as it was a human worker, bondsman, or slave who was charged with the veiled operation of the centre.[25] However, 'man cannot leave the center of operation in the dark, when he no longer intervenes as tool bearer; it is the center that must effectively be produced by the technical object'.[26] Simondon thus argues that the philosophical doctrine of hylomorphism is essentially subject to challenge from the material reality of the technical object rather than from a socio-political or philosophical critique of the master-slave dynamic at work in hylomorphism. While in artisanal, pre-modern times, it was indeed the human who was responsible for the mediation between form and matter, with the advent of the technical machine taking over the responsibility of mediation, the representation of the 'way of functioning that coincides with technical operation, which accomplishes it'[27] has necessarily been put into question.

The underlying argument here is, then, that insight into technical operation proper could remain structurally unquestioned until the invention of post-artisanal technical objects. This is the case because, in taking over the worker's role of mediating between form and matter, the industrially produced technical object, the machine, necessarily subjected the veiled centre to inquiry. For Simondon, there is then a specific historical moment that brought about both the possibility and necessity of developing post-metaphysical, technical thought proper. This specific historical moment is the transition from the artisanal to the industrial mode of production, with its industrially produced technical objects and machines. The conditions for technical knowledge proper and post-metaphysical thought are said to be the result of the industrially produced technical object.

This direct interference of the industrial technical object on the development of technical and philosophical thought is, however, only one side of Simondon's account of the history of technical thought. The underlying argument of *On the Mode of Existence of Technical Objects* is that to this day, and thus long after the invention of the industrial technical object, knowledge of technical objects remains 'nonessential' in character nonetheless.[28] While the material conditions for both 'essential' technical knowledge and post-metaphysical philosophical thought have been in place since the Industrial Revolution, these conditions have yet to be actualized. Simondon's project articulated in *On the Mode of*

Existence of Technical Objects is thus twofold in nature. First, Simondon argues that culture 'has constituted itself as a defense system against technics; yet this defense presents itself as a defense of man, and presumes that technical objects do not contain a human reality within them'.[29] It is thus the exclusion of technics from culture proper, which is to blame for the non-actualization of technical knowledge proper, giving rise to prevalent tensions and conflict, anxiety, alienation, and resentment in the relation between technics and the human. Simondon's explicit aim, then, is to first rediscover this alienated human reality at the heart of the technical object and to integrate it into culture in order to resolve present social conflicts and contradictions. Second, in sketching out the role that technical operation and the Industrial Revolution play for the development of technical thought and, in a second step, also philosophical thought, Simondon aims to at last introduce and actualize the dormant conceptual resources that have materially been available since the Industrial Revolution. In so doing, he seeks to overcome the metaphysical limits to knowledge and to finally put forward a post-metaphysical philosophy – his very own philosophy of individuation – capable of doing justice to the mode of existence of technical objects.

Simondon's history of technical thought

Thinking back to the discussion of Heidegger, it thus becomes clear that when it comes to the history of technical thought, Simondon tells a radically different story. While Heidegger understood modern metaphysics as the movement from the objective mode of existence to the standing reserve (*Bestand*), in the sense that metaphysics in its final stage becomes modern technics, Simondon instead argues that metaphysics was founded on an incomplete technical experience, according to which, since the very beginning of metaphysics in ancient Greece and onwards, technical experience is reduced to work. Furthermore, Heidegger proposed Kant's *Transcendental Analytic* as the moment in which modern metaphysics becomes technics, arguing that the nature of his own time was the final coming-to-end of metaphysics. Simondon, on the other hand, claims that it is in fact the industrially produced technical object, in the form of the machine itself, that brought about the conditions of possibility for post-metaphysical thought and technical knowledge proper. While both locate this moment of transition at the end of the eighteenth century, the challenge to the metaphysical conception of technics was posed, according to Simondon, by the material technical object itself. What both thinkers share is an agreement that the end of metaphysics is yet to be brought about and an explicit aim to end it, either through the process of questioning (Heidegger) or problematization (Simondon).

The preceding discussion has shown how Heidegger positioned Newton, Kant and Nietzsche between Descartes and his own historical present, none of whom Simondon explicitly engages with at length. However, both Newton and Nietzsche fit into the Simondonian narrative without great difficulty. Newton is, without doubt, the most famous representative of what Simondon calls the properly universal science of technical *Encyclopedism*, while Nietzsche's will to power is what Simondon terms the 'autocratic philosophy of technics'.[30] The one thinker that appears to fit into the Simondonian narrative with greater difficulty is Kant, who is explicitly mentioned only once in *On the Mode of Existence of Technical Objects*, in reference to his moral and political philosophy.[31] But, as Muriel Combes first argued and which I will myself show in the following, Simondon's technical thought is at once explicitly aimed against Kant's Transcendental Philosophy, while it at the same time pursues, what is without doubt, a quasi-Kantian project of its own.

Simondon and Kant

In her widely read book *Gilbert Simondon and the Philosophy of the Transindividual*, Muriel Combes claims that Simondon's philosophy of individuation is explicitly aimed against 'theories of knowledge inspired by Kant'.[32] While Kant famously set out to establish the a priori conditions of and thus the possibility and the limits of knowledge, Simondon argues that such a priori cognitions of conditions are impossible, as 'we cannot *know individuation* in the ordinary sense of the term; we can only individuate, be individuated, and individuate within ourselves'.[33] As such, Simondon's philosophy of individuation explicitly rejects the a priori distinction between the subject and object of knowledge which he considers to be intrinsically hylomorphic, as it operates from the standpoint of the difference between a priori forms of cognition and the a posteriori matter of sensibility. Consequently, for Simondon, Transcendental Philosophy is either too early (a priori) or too late (a posteriori), but never contemporaneous with the becoming of technics. From the perspective of Simondon's project, technics is thus the one question that Kant is structurally incapable of answering to. His answer will either come too early or too late.

Consequently, and from a structural perspective, Simondon's project is oriented against Kantianism in all its forms. While there is a surprising absence of explicit references to Kant in Simondon's writings, one must not, at the same time, overlook Simondon's simultaneous proximities to the Kantian philosophical project. Looking into the Simondonian account of the history of technical thought, we can find an important conceptual distinction that does not feature in either Heidegger's or Stiegler's accounts. The technical object, Simondon claims, can be related to the human in two fundamentally different ways, both of which continue to exist up to the present day but were dominant at different points in time. On the one hand, technics can be encountered and acquired non-reflectively by the human during childhood, giving rise to a relation of 'minority', and on the other hand, it can be reflectively learnt during adulthood, giving rise to a 'majority' relation between the human and technics. What becomes clear, then, is that Simondon engages a distinction famously employed by Immanuel Kant in his 1784 Prize-essay 'An Answer to the Question: "What Is Enlightenment?"' Enlightenment (*Aufklärung*) was principally defined by Kant as the 'human being's emergence from his self-incurred minority',[34] with minority referring to a difficulty or inability of judging without the help of others. While only exceptionally achieved by individual people, Kant's argument ultimately aims to locate the self-emergence from minority in the public as a whole and in its free use of reason by means of the instrument of criticism.[35]

Aligned with influential critiques of Kant's Enlightenment discourse put forward by thinkers including Nietzsche, Adorno and Habermas, Simondon at once employs Kant's Enlightenment discourse and, at the same time, attempts to turn it against itself, since what is at stake, according to Simondon, is not pure theoretical or practical reason but rather technical reason. The minority relation between technics and the human laid open by Simondon determines the technical object as 'firstly an object of utility, necessary for everyday life, belonging to the heart of the environment where the human individual's growth and training takes place. In this case, the encounter between the technical object and man occurs essentially during childhood. Technical knowledge is implicit, non-reflective, and habitual'.[36] Built by habit and thus functioning as a quasi-second nature, the minority relation relies on having been taught during childhood, making it difficult, if not impossible, to self-reflexively change one's relationship to the technical object after childhood. The knowledge that is said to ensue from the minority relation is thus 'operational rather than intellectual; it will be an ability rather than knowing; by its nature, it will be a secret for others, because it will be a secret for himself, for his own consciousness'.[37] Essentially lying outside of conceptual and scientific discourse, there is as such no, or at least no direct, relation between science and technics possible from within the minority relation. The minority relation to technics is, according to Simondon, lived by miners, shepherds, farmers, artisanal technicians in general, and so on.

Complementing the minority relation between technics and the human, however, Simondon puts forward his account of a 'majority' relation to technics, an 'enlightened' technics, which 'corresponds to an operation of reflection and self-awareness by the free adult, who has at his disposal the means of rational knowledge elaborated through the sciences: the knowledge of the apprentice is thus distinguished from that of the engineer'.[38] While technical minority is acquired during childhood, technical majority is the technical relation acquired by the human during adulthood, it corresponds to the rational, scientific and thus universal insight of a studied engineer:

> It is rational because it employs measurement, calculation, procedures of geometrical figuration and descriptive analysis; [. . .] not only is scientific explanation required, but it is required with a clear taste for the scientific spirit. Moreover, this way of teaching is doubly universal, both through the public it addresses and through the information it provides.[39]

Simondon's example of the majority relation between technics and the human is Diderot and D'Alemberts's *Encyclopedia*,[40] published between 1751 and 1772 in France. Containing prints of schemas and models so that its readers could both build as well as develop the depicted technical apparatus further by means of invention, Simondon argues that the *Encyclopedia* first freed technical knowledge from the protection of the guild,[41] and second made it accessible to the wider public for the first time. At the same time, however, the *Encyclopedia* provides more than just the prime example of the majority relation to technics. Simondon's argument is that, in fact, the majority relation as a whole is itself the effect of what he calls encyclopedic spirit, or *Encyclopedism*.

Simondon argues that the first appearance of *Encyclopedism* emerged as early as the Renaissance,[42] when technical machines were first introduced into the realm of rationality and science.[43] Ultimately held back by the immaturity of the sciences, Simondon reads the eighteenth-century Enlightenment as the second stage of *Encyclopedism*, a technical *Encyclopedism* which properly 'freed' technical thought for the first time.[44] Diderot and D'Alembert's *Encyclopedia* was thus at heart a book moved by this technical spirit or force that was first freed by Enlightenment science.[45] This technical force, Simondon writes, 'existed by itself, because it responded to the needs of its time, more than political or financial reforms did',[46] realizing nothing less than a society of technical autodidacts who could think for themselves and bring forth technical objects by themselves. Simondon thus argues that it was technical *Encyclopedism* that brought about the essential opposition to the 'tutelage and the status of spiritual minority', culminating in the moving forces that would finally overthrow the 'moral and political heteronomy of the monarchy [*l'Ancien Régime*]'.[47] Understanding the second stage of *Encyclopedism* to have been a properly technical *Encyclopedism*, Simondon thus argues that it was ultimately 'through technics'[48] that the eighteenth century managed to free itself from the social, moral and political constraints of *l'Ancien Regime*.

Simondon's periodization of *Encyclopedism* reads the history of modernity through a technical lens, going as far as to argue that Enlightenment humanism was at heart a technical force of liberation. Furthermore, however, Simondon's periodization of *Encyclopedism* itself centres on the process of convergence between science and technics/technology. While the first introduction of technics into science is said to have taken place already during the Renaissance, it was only the enlightened stage of *Encyclopedism* that finally 'freed', in the sense of universalized, technics, making possible, for the first time, a majority relation to technics. This second, enlightened, and properly technical stage of *Encyclopedism* was itself, however, held back by 'the hierarchical aspects of social rigidity',[49] Simondon writes. What is needed in order to complete the convergence between science and technics, then, is a third type of *Encyclopedism*, a technological *Encyclopedism*.

The translators of *On the Mode of Existence of Technical Objects* inserted a note into the *Introduction* in order to highlight the difference between technics and technology. While technics (*la technique, les techniques*) refers to both 'the practical and particular application of technics to a given concrete task' as well as 'the theory or study of industry and of the mechanical arts', technology, for Simondon, constitutes the philosophical meta-theory of technics.[50] While Simondon explicitly discusses his notion of technology in only a few passages, the entire project of *On the Mode of Existence of Technical Objects* can be read as Simondon's attempt to produce just such a philosophical meta-theory of technics. In fact, for Simondon, the duty of such a philosophy of technology is 'analogous to the one [philosophy] fulfilled for the abolition of slavery and the affirmation of the value of the human person'.[51] This means that, just like its two predecessors, technological *Encyclopedism* remains to be at heart a humanism aimed at liberation.

Differently from the enlightened humanism of the eighteenth century, however, Simondon's claim is that the twentieth century must seek 'a humanism capable of compensation for the form of alienation that intervenes within the very development of technics, through a series of specializations that society demands and produces'.[52] The specific form of alienation that technological *Encyclopedism* is tasked with liberating is, according to Simondon, explicitly technical. For Simondon, what is needed in his day and age is a humanism that recognizes and becomes aware of the mode of existence of technical objects in order to free 'human reality enclosed within the technical object'.[53] Such a liberation cannot be achieved by a repetition of the universalizing efforts of the enlightened, technical *Encyclopedism*. And furthermore, it is no longer a problem of merely putting science and technics in relation to one another. If technological *Encyclopedism* is to tackle the issue of technical alienation, it first needs to introduce 'a representation and scale of values adequate to the essence of technical objects'[54] into culture in order to then, in a second step, develop a universal symbolism common to both the human and the machine. This is a twofold effort that Simondon first finds in Norbert Wiener's *Cybernetics* (1948), as well as subsequent theorizations of cybernetics.

As the preceding pages have shown, Kant resurfaces at several moments as Simondon's silent interlocutor in his discussion of *Encyclopedism*. Simondon's distinction of the human's two possible relations to technics are furthermore undoubtedly references to a terminology famously at play in Kant's Enlightenment essay. From the conceptual distinction between technical minority and majority (*minorité* and *majorité*),[55] to the terms of tutelage and spiritual minority,[56] one can observe Simondon continually referencing Kantian terminology. However, while explicitly employing Kantian terminology, Simodon's implicit critique is that Kant could not see, as he was structurally incapable of it, that the driving force of the Enlightenment was nothing but technical. Kant's problematic reaffirmation of the hylomorphic matter/form distinction that had dominated philosophy ever since ancient Greece means for Simondon that, despite writing at the time of *technical Encyclopedism* and the advent of the industrially produced technical object, Kant remained structurally incapable of thinking technics, even, or especially, in its instantiation as universal, technical reason. In a surprisingly similar fashion to Heidegger, according to whom Kant's *Transcendental Analytic* was nothing but a *Transcendental Technics*, in Simondon's reading, the *Critique of Pure Reason* as a whole could or rather should have been called *Critique of Technical Reason*. The fact that it was not means that technicity remained unaccounted for and thus alienated within the efforts of Enlightenment humanism. It is this very occlusion that demands yet another *Encyclopedism*, a technological *Encyclopedism*, capable of at last attending to the mode of existence of technical objects proper. On the one hand, then, Simondon reads the Enlightenment and the Industrial Revolution as two historical moments that brought about the conditions for both scientific technical thought proper and post-metaphysical thought. But, much like Heidegger, Simondon's claim is that these conditions remained unactualized and that it will be his own contribution to technical thought as well as to philosophy at large, which will finally actualize the conditions for post-metaphysical technical thought proper that have been dormant for the past, approximately 150 years.

3

Stiegler and the repression of technics

If Heidegger argued that the right question concerning technics has not been asked prior to his own intervention, and if Simondon claimed that metaphysics was out of time with technics in its problematic becoming, then Bernard Stiegler repositioned the narrative of philosophy's technical aporia as a problem of repression. *Technics and Time, 1: The Fault of Epimetheus* (*La technique et le temps, 1: La faute d'Epiméthée*), forming the first installment of his *Technics and Time* series, was published in 1994 in the wake of deconstruction and at the time of a French resurgence of interest in Simondon's work. Following Simondon and Heidegger, Stiegler prefaces his own philosophical intervention by articulating the underlying technophobic constitution of philosophy. And thus, the book opens with a familiar refrain according to which technics 'is the unthought'.[1] A new undertone can however be detected in Stiegler's particular reiteration. The problem here does not seem to be one of living in a technological world devoid of adequate conceptual tools for addressing it. It is not that philosophy is out of time with technics, or that it simply forgot about technics. Rather, Stiegler explicitly frames the relation between philosophy and technics as one of repression, whereby echoing Derrida's articulation of the privileging of speech over writing as an act of repression in *Of Grammatology*.[2] Employing this Derridean framing, Stiegler then writes that 'at its very origin and up until now, philosophy has repressed technics as an object of thought'.[3] Philosophy, he thus claims, has repressed technics since its very beginning, meaning that technics 'is not un-identified in the sense in which something forgotten is not thought: it is largely thought and felt to be unthinkable'.[4] By framing technics as the 'unthought' of Western thought,[5] Stiegler first proposes that, to attempt to think technics demands a quasi-psycho-analytic mode of inquiry about thought, as proposed by Tania Espinoza.[6] Second, Stiegler stages his own intervention into the history of technical thought as an act of liberation, in which technics is finally to be set free. As I will show, Kant will once again come to play an important role within Stiegler's liberation narrative. While explicitly performing a thorough critique of Transcendental Philosophy, Stiegler's engagement with Kant did not only effect a restructuration of the entire *Technics and Time* series, it also allowed Stiegler to overcome a 'connective fault'[7] at the core of his project. In the following, I will first read Stiegler on his own terms, before tracing his account of the history of technical thought and the role assigned, as well as denied, to Kant in this history.

Introducing Stiegler's framework of exhumation

Stiegler's project of the liberation of technics proceeds by way of cross-reading twentieth-century French paleanthropology on the one hand and ancient Greek mythology as philosophy's pre-history on the other, so as to critique and reshape the canonical texts through which technics has hitherto been determined. First, and following the French anthropologist André Leroi-Gourhan who famously put forward the thesis of the 'originary characterization of the anthropological by the technological',[8] Stiegler argues that technics is the originary prosthetic supplement of the human. What is specific to Stiegler's claim here is that this supplement does not replace something that has gone astray and is now lacking. The argument is rather that the human is in 'default of origin',[9] in the sense of originarily being without either quality or predestination.

The default of origin is, in a second step, further determined by way of the ancient Greek myth of Prometheus and Epimetheus, which Stiegler reads through Jean-Pierre Vernant's interpretations of the myth in Plato and Hesiod.[10] According to the myth, Prometheus' brother Epimetheus forgot to give an attribute to the human when charged with the task of distributing qualities to all beings. In order to make up for the initial fault of his brother, Prometheus engages in the cunning theft of fire from Hephaestus so as to provide the human with a prosthetic supplement after the fact, thereby doubling up on the initial fault of Epimetheus. The concept of prosthesis developed by Stiegler engages the human in a threefold relation. It firstly establishes a spatial relation in the sense that the human is placed in front and outside of itself: 'In order to makeup for the fault of Epimetheus, Prometheus gives humans the present of putting themselves outside themselves'.[11] At the same time, prosthesis also establishes a temporal relation in a double sense. The prosthetic both sets in advance – in the sense of what lies in the past – as well as giving the human the capacity of anticipation and foresight, and thus its relation to the future as its ultimate possibility.[12] In this temporal relation, Stiegler argues that technical prosthesis functions as a special kind of memory, which he calls epiphylogenesis.[13] Epiphylogenesis designates the accumulation of individual experiences and traces inscribed and collected in technical artefacts, through which they can be passed on through time. Stiegler draws his notion of epiphylogenesis from Leroi-Gourhan's threefold distinction of memory and replaces the notion of technological memory[14] that he finds in Leroi-Gourhan with his own epiphylogenesis. Governed by technico-social rather than genetic differentiation, the technical artefact, for Stiegler, functions as an external memory support of a past that none of us have lived but which we have inherited and are to adopt as 'our' own: it is our 'already there'. Performing a powerful critique of Heidegger's notion of the 'already there',[15] Stiegler then argues that Heidegger has forgotten the originary prostheticity of the 'already there' that is essentially constitutive of *Dasein*.[16] The implication of Stiegler's point is indeed powerful, as it follows that despite Heidegger's technical critique of Kant's *Transcendental Analytic* and in moving from Kant's *Transcendental Analytic* to his own *Existential Analytic* in *Being and Time*,[17] Heidegger has himself forgotten the technical prostheticity constitutive of *Dasein*.

Stiegler's account of technics as a prosthetic external memory support in essence puts forward a quasi-transcendental account of technics. As both in front and outside of 'us' while also being our 'already there', the technical prosthetic mediates our relation to the world in the sense of first giving us access to the world in the sense of a proposition.[18] Technics, Stiegler writes, 'is what is placed before us [*la technique est ce qui nous est pro-posé*] (in an originary knowledge, a *mathesis* that "pro-poses" us things)'.[19] Stiegler can thus be seen to at once follow Heidegger's reading of *mathesis* as enframing (*Gestell*) while at the same time pushing Heidegger's thought outside of itself. For Stiegler, technics *is* transcendental *mathesis*, an 'originary' knowledge that opens us onto the world. At the same time, however, the concept of epiphylogenesis essentially suspends the very distinction between the transcendental and the empirical. The prosthetic, always encountered in the empirical world and thus in a sense a posteriori, at the same time precedes 'this consciousness in time as the possibility of its already-there',[20] revealing the a priority of the transcendental to be a strange after-effect of the prosthetic.

First and straightforwardly, Stiegler thus argues that both the metaphysical opposition between *logos* and *techne* and the natural-scientific qualification of technics in opposition to both the organic and the inorganic fail to be useful. What follows from this first aspect is immense, for if technics is elevated to the status of an originary knowledge and rearticulated as a quasi-transcendental, it becomes clear that 'religion, speech, politics, invention – each is but the effect of the default of origin'.[21] According to Stiegler, *logos* and *techne* are thus both modalities of the originary default of the human, in relation to which they engage in a composition rather than an opposition. They are both, Stiegler writes, 'the fruit of [humanity's] incompleteness'.[22] And from this point on, technology is hyphenated.

There is, however, a further point to be made regarding Stiegler's argument on the composition between *logos* and *technics*, which concerns their respective statuses of autonomy and heteronomy assigned by metaphysics. Traditionally, the noetic thought of *logos* is said to be self-moving and autonomous, while technics, since Aristotle, is instead posited as moving heteronomously. This is due to the fact that ever since Aristotle, *techne* is argued to lack *arkhe*, that is, the principle of its own movement and rest. Consequently, the technical prosthetic ever since Aristotle has been subject to a metaphysical denigration, reduced to a mere means devoid of its own proper dynamic. Leroi-Gourhan's quasi-zoological investigation into the development of technical artefacts, and Simondon's concept of concretization allow Stiegler to powerfully contest this denigration. Technics, Stiegler concludes, possesses a universal tendency driving its evolution from within, a point with which Stiegler contests the Lamarckian natural-scientific distinction of beings into two classes. From the Lamarckian biological perspective of the twofold distinction into the organic and inorganic, 'technical things' are left hanging somewhere in the middle, suspended as a quasi-monstrous hybrid.[23] Once a universal technical tendency is at play, however, technical things can neither be said to be organic and thus self-organized, nor can they any longer pass as unorganized and inorganic matter. Instead, Stiegler claims that such a technical tendency necessitates the introduction of a third kind of being, which, as it stands in a particular and originary relation to the human as 'the pursuit of life by means other than life',[24] he calls 'organized inorganic matter'.[25]

Stiegler's account of the history of technical thought

Stiegler's liberation of the concept of technics has so far been seen to proceed by way of a transdisciplinary determination, at once critiquing and aiming to reshape the Western philosophical canon held to be responsible for the repression of technics up until Stiegler's own intervention. Complementing his reading of paleoanthropology, Stiegler's engagement of the Promethean myth

stages a return to the 'beginning' of the Western philosophical canon, that is, the moment of philosophy's self-constitution via the repression of technics in classical Greece. 'At the beginning of its history philosophy separates *tekhne* from *episteme*, a distinction that had not been made in Homeric times. The separation is determined by a political context, one in which the philosopher accuses the Sophist of instrumentalizing the *logos* as rhetoric and logography, that is, as both an instrument of power and a renunciation of knowledge.[26] In an attempt to cleanse itself from the danger of instrumentalization, philosophy thus pitches scientific knowledge against technical knowledge, by way of which technical knowledge will become disparaged for several centuries to come. Stiegler then points out that the issue with the Platonic resolution to the problem of instrumentality is that it is based on a fundamental misinterpretation. If *logos* is indeed in danger of being instrumentalized, then the possibility of instrumentalization is inscribed in *logos* itself rather than being an effect of a corruption by *technics*. The fundamental problem, then, is not the relation that *logos* does or does not have to instrumentality, nor is it instrumentality itself. Rather, Stiegler locates the fundamental problem in the misconception and consequent reduction of the instrument 'to the rank of means',[27] resulting in the subsequent attempt to distance oneself from the instrumental together with the political intention to master it.

Stiegler's brief account of philosophical modernity begins with the seventeenth century and Descartes, who posited the 'I think' as the constituting subject that faces an object constituted by the subject and which is, in a further step, to be mastered by the subject through the instrument of technics. But while this new consciousness of the 'I think' had been in effect since the seventeenth century, Stiegler's claim is that it was only fully 'authorized[28] by the subsequent technoscientific modernity that set in with the nineteenth century, characterized by the thermodynamic and industrial revolutions. Stiegler's claim about the relation between philosophical and technoscientific modernity is complex. First, and in reference to the Enlightenment project of public education and discourse, his claim is that the 'I think' and its complementary discourse of technical mastery over nature were 'concretized and generalized during the nineteenth century at the heart of the first Industrial Revolution'.[29] As such, his argument is here one of historical continuity between the philosophical invention of modern consciousness and its large-scale implementation two centuries later, giving rise to the technical consciousness that would essentially drive technoscientific modernity. Alongside the role played by public education, the full authorization of modern consciousness also stands in an important relation to the thermodynamic concepts of energy and metastability, thoroughly unsettling and in fact reversing the old metaphysical order according to which stability was the rule and change the exception.[30] It was thus not until the thermodynamic claim of a thoroughly unstable world in which change is the norm, that the idea of an all-powerful subject, who ventures to transform this formless, chaotic nature, was brought full circle.[31]

Stiegler follows up his argument on the continuity between the philosophical invention of modern consciousness and its large-scale implementation two centuries later by arguing that the same technoscientific modernity which first completed philosophical modernity would also lead to the eventual break with the modern subject position and the accompanying understanding of science and technics that has been in place from Descartes until Kant. Stiegler's argument here is that the experience of technoscientific modernity, of living in a technological world in which technics seems to have gone out of control and can no longer adequately be understood as the application of science revealed – to the 'naked eye[32] – the problems with the modern understanding of technics. Stiegler's reading of the history of technical thought is here supported by a phenomeno-materialist argument, which one also finds in the work of Simondon, according to which it was the invention of modern, technical machines and the phenomenological experience of living among those machines that demanded the elaboration of a new thought of technics.[33] As Stiegler writes, 'since the Industrial Revolution, "technical becoming" [. . .], has compounded its systematic dimensions, becoming visible to the naked eye in various ways and sensible to the bodies and minds devastated by an entire universe of hellish machines.'[34] Stiegler's point is thus that prior to the experience of the industrially produced technical system, the inventiveness, evolutionary logic, and systematic dimension characteristic of technics proper had not been thinkable. For Stiegler, it is thus the industrially produced technical object, which is the condition of possibility of a 'new' technical thought. Consequently, Stiegler argues that it was only in the nineteenth century that the possibility of a 'techno-logy that would constitute a theory of the evolution of technics',[35] first became possible.[36] This new technical thought is conceptualized by Stiegler as the after-effect of the phenomenological experience of living among 'hellish machines'. What follows from this first *opening* of the possibility of technology is, however, first a nihilist techno-pessimistic discourse from Nietzsche's will to power, to Husserl's *Crisis of the European Sciences*, until Heidegger's account of technics as *Gestell*, all of which broke with the historical link between technics and objectivity, as well as the modern understanding of science, put in place by philosophical modernity.[37]

There is a second moment in Stiegler's narration of the history of technical thought which carries a distinct Simondonian tone. In *Technics and Time 3*, we find Stiegler arguing that there was yet another 'unique event',[38] a historical event, which revealed the true structure of temporal objects and, as a consequence, of consciousness itself.[39] The unique event that Stiegler is referring to is the invention of a specific technical object, the phonograph, by Thomas Edison in 1877, which would later be called the Gramophone. Stiegler here thus argues in line with his earlier argument, according to which the industrially produced technical object brought about the condition of possibility of technical thought proper, that the condition of possibility for philosophical insight into the proper workings of consciousness are technical and were actualized through the technical invention of a specific technical object called the phonograph. What was made possible by the invention of the phonograph is nothing less than the 'phonographic revelation of the structure of all temporal objects'.[40]

The second periodization category employed by Stiegler is that of the 'old metaphysical doxa',[41] which, according to Stiegler, spans the entirety of the history of Western philosophy that opens with Aristotle and closes with Kant. Stiegler here writes that 'at least from Aristotle to Kant, technics [. . .] arises from neither the practical domain as such nor the theoretical domain, in which it is cancelled'.[42] He then goes on to insist that 'no dynamic *proper* exists for Aristotle, any more than for any other metaphysician – nor thus for Kant: this is their common feature'.[43] What is shared by Aristotle and Kant is thus an understanding of technics in terms of a means/ends rationality devoid of the systematicity of science and devoid of a universal tendency driving technical evolution from within. Given their vastly different historical contexts, Stiegler points out that Aristotle and Kant are nevertheless differentiated by the modern concept of science. While both Aristotle and Kant consider science to be 'what announces and formalizes the real *as what cannot be otherwise*',[44] modern science sees in technics an 'application of science',[45] while, for Aristotle, technics was constituted in opposition to science.[46]

In a problematic move on the part of Stiegler himself, his designation of the second periodization as the 'old metaphysical doxa' must be seen to reduce practically all of the history of philosophy, from Aristotle to Kant, to a problem of conjecture and belief.[47] Doxological critique was historically employed by Plato in order to separate *logos* from its instrumentalization by the sophists.[48] In a strange turn of events, Stiegler himself employs the metaphysical distinction between philosophy and sophism, or science and technics, that he himself aimed to displace otherwise. Moreover, it performs a powerful function within

Stiegler's account of the history of technical thought. Essentially, Stiegler argues that this 'old metaphysical doxa' came to a close after Kant, whereby he differentiates his account of the end of metaphysics from that of both Heidegger and Simondon. As the previous section has shown, Heidegger located his own intervention at the long tail end of the completion of metaphysics, which is to be brought about by his discourse of questioning, while for Simondon the Industrial Revolution surely produced the conditions of possibility for a non-metaphysical thinking of technics, but due to culture's technophobic constitution, it failed to bring about genuine technical insight. The actualization of this post-metaphysical thought on technics is consequently the subject of Simondon's own philosophical contribution in *On the Mode of Existence of Technical Objects.*

Stiegler's account of the history of technical thought can be seen to move through the history of European philosophy in broad strokes, beginning from the argument that after philosophy's self-constitution by way of the repression of technics in ancient Greece, it was not until the thermodynamic and industrial revolutions of the nineteenth century that contemporary thought on technics proper first became possible. And while Marx first opened the possibility of theorizing the systematicity and evolution of technics, Heidegger still forgot about the technical prostheticity constitutive of *Dasein* seventy years later in his *Being and Time*. In his own project of, at last, exhuming/inventing technical thought proper, Stiegler had initially envisioned his *Technics and Time* project to encompass three volumes in total. Upon embarking on the third volume, he encountered a problem of connection, a 'connective fault'.[49] This problem of connection, he goes on to argue, could only be resolved through an in-depth engagement with the very 'heart of modern philosophy': Kant's *Critique of Pure Reason*. Stiegler explains that it was his engagement with Kant which finally allowed him to recognize and work through this 'connective fault', that is, how to bring into view 'the focal point of the very idea that, despite many attempts had escaped me [Stiegler]'.[50] Following a reading of the *Critique of Pure Reason*, Stiegler thus decided in the year 2000 to push back what was initially supposed to have been the third and last volume of the *Technics and Time* series to a fifth position, inserting new third and fourth volumes. How, then, could Kant, who in the preceding discussion was shown to occupy the problematic closing positions in both the categories of philosophical modernity and the old metaphysical doxa before the conditions of possibility for technical thought proper were even opened, possibly help Stiegler in his project of the liberation of technics proper?

Stiegler and Kant

The third volume of *Technics and Time* occupies a convoluted position in relation to the overall series. Structurally speaking, there are two aspects to Stiegler's engagement with Kant. First, and most importantly, Stiegler's engagement with Kant demanded, as we have seen, nothing less than the restructuring and overall rethinking of the entire *Technics and Time* series, which now spans a total of five rather than merely three volumes. Second, however, Stiegler's reading of Kant now explicitly prepares the ground for and serves as the introduction to the entire *Technics and Time* series. Essentially performing a reading of Kant, *Technics and Time 3* is said to be both autonomous of the series while at the same time functioning as an introduction to the preceding two volumes.[51]

On the level of conceptual engagement, Stiegler's reading of Kant's *Critique of Pure Reason* centres on four main points: synthesis, schematization, orientation, and finally critique. Of these four issues, it is the question of synthesis that constitutes 'the heart of the reflections [Stiegler] offer[s] here through a reading of Kant's *Critique of Pure Reason*'.[52] In line with Stiegler's argument, presented in the preceding pages, that the a priori is essentially the after-effect of prostheticity, Stiegler accuses Kant of not having seen how, behind the a priori threefold synthesis of consciousness looms a fourth synthesis. This fourth synthesis, Stiegler argues, is technological and synthetic, in the sense in which 'we call the artifice of prosthetic replication "synthetic"'.[53] Stiegler's point is thus that Kant himself always already relied on a number of technical 'retentional instruments',[54] and that it is precisely due to these material, technical traces that Kant's conscious activity first became accessible to both himself and his public. Functioning as a prosthetic milieu, at once consciousness and its other,[55] Stiegler argues that the unity of Kant's own thought is only accessible to Kant through his books, notes and other technical traces, functioning as the understanding's 'veritable crutch'.[56] While Kant always relied on this fourth, technological synthesis, he nevertheless remained incapable of seeing, understanding, or thinking it in his transcendental idealist insistence on the interiority of phenomena.

Stiegler's second point of engagement with Kant sets off from Adorno and Horkheimer's critique of the culture industry in their 1944 *Dialectic of Enlightenment*. Their critique of the culture industry and specifically Hollywood cinema hinges on the claim that its 'prime service to the customer is to do his schematizing for him'.[57] Hollywood cinema is here said to have finally deciphered and gained control over what, according to Kant, was a 'hidden art in the depth of the human soul'.[58] Stiegler's engagement with Adorno and Horkheimer's argument on schematization asks how it is possible to schematize for someone else. Essentially employing the Kantian methodology of inquiring into the conditions of possibility while at the same time arguing against Kant, Stiegler's claim is that it was Kant who did not acknowledge the primordial role that mnemo-technical retentions (technical objects) have always already played in the constitution of consciousness. Stiegler thus ventures to show how image and schema, material trace and concept are coemergent and interrelated, the reality of which is said to have escaped both Kant as well as Adorno and Horkheimer. Neither of them recognized the technical 'substratum'[59] of the 'third' provided by the schema. For Stiegler, the schema and its substratum are nothing but 'two faces of the same reality',[60] providing the twofold after-effect of an onto-epistemological elaboration that takes place prior to the bifurcation into form and matter, the inner and outer, the mental and the material. What this means, then, for Stiegler is that there can only be such a thing as an industrial schematism 'because the schematics are originarily, in their very structure industrializable: they are functions [. . .] of technics, technology, and, today, industry'.[61]

Stiegler's third and fourth points of engagement with Kant equally do not stop short of being thoroughly unsettling. In naming the *Critique of Pure Reason* 'Critique', without a grammatical proposition that would concede the possibility of there ever being other possible critiques of pure reason, Kant put forward the performative claim to have completed the project of reason's critique once and for all. However, in his discussion of the current technoscientific paradigm, Stiegler joins in the choir of post-Kantian thinkers who challenge the Kantian position. Stiegler's particular reiteration is that when it comes to technoscience, by which he means contemporary scientific practice, the relation between the real and the possible has essentially been reversed. This is the case because technoscience, rather than describing or accounting for an already existing reality, aims to 'create a new reality'.[62] As such, the Kantian founding distinction between theory and practice, including the restrictions put on theoretical reason, is not only transcended but essentially invalidated. The questions at stake in Stiegler's discussion of technoscience first point to the necessity of a new, post-Kantian critique informed by technics, which, second, asks about the conditions of possibility of judging the technological fictions produced by technoscientific practice as a problem of how to orientate oneself in the 'darkness of scientific possibles'.[63]

Throughout this four-partite, critical discussion of Kant, Stiegler refers to Kant's explicit writings on technics only once in order to underscore his point that Kant indeed mistook the relation between science and technics as a problem of mere application.[64] Concluding that Kant had little to say about technics himself, Stiegler then posits Kant as the closing figure of both his periodization categories preceding the possibility of technology. At the same time, however, the preceding discussion has shown just how important Kant was for Stiegler's own technics-thinking. It was Stiegler's engagement with Kant that first revealed to Stiegler the long-sought 'connective fault'.[65] And it was Kant who, in the following, catalysed the complete restructuring of the *Technics and Time* series, with the third book containing Stiegler's reflections on Kant being repositioned as the introduction to the entire newly restructured series. Thus, similarly to the preceding discussions on Heidegger and Simondon, Kant occupies once again a twofold position. At once denied insight into technics, Kant at the same time prepares the ground and catalyses Stiegler's own intervention.

Concerning Stiegler's engagement with Kant on the level of conceptual content, one is reminded of the preceding symptomatic readings of Kant once again. In Stiegler's account, Kant is read as the last pre-industrial thinker, who so powerfully repressed the technical conditions of possibility for both synthesis, schematization, orientation, and critique in the last instance. Taking Stiegler's discourse of technical repression to its logical consequences, it indeed appears that Kant could not 'know' anything about the technical constitution of schematization. According to Freud, forgetting accompanies repression, due to which 'the motives for the prohibition (which is conscious) remain unknown; and all attempts of disposing of it by intellectual processes must fail'.[66] And indeed, when faced with the question of laying open the mysterious process of schematization, Kant could only characterize it as a 'hidden art in the depth of the human soul'.[67] It is this very unthinkability of technics, which Stiegler ultimately has in mind, when he writes that once repressed, technics 'is not un-identified in the sense in which something forgotten is not thought: it is largely thought and felt to be unthinkable'.[68] Consequently, then, while catalyzing Stiegler in articulating the mnemo-technical nature of schematization, Kant is at the same time not read as a thinker of technics. Instead, Kant is read by Stiegler as having 'unthought' technics in all of its aspects; heavily relying on mnemo-technical prosthetics, technics is said to have remained an implicit object of thought for Kant. Thus, in a similar way to Heidegger, Kant is read by Stiegler as conditioned by and thus ultimately suffering from rather than thinking technics, as well as philosophy's age-old repression of it. It is not until his own intervention that technics, according to Stiegler, is liberated at last. However, given the near absence of consideration afforded to any explicit thinking on technics by Kant, it remains unclear what exactly Stiegler thought that technics could have meant for Kant himself. As with the preceding discussions on Heidegger and Simondon, Kant occupies what can only, once again, be said to be a paradoxical position. Kant is at once denied insight into technics, while at the same time preparing the ground for Stiegler's own intervention.

4

A future for Kant despite Heidegger, Simondon, and Stiegler

The preceding discussions have shown that and how Heidegger, Simondon, and Stiegler argue, in one way or another, that technics articulates the limits of philosophical thought understood as metaphysics. Technics is at once philosophy's fundamental obstacle and its possibility. If philosophy is to finally open itself up to technics, it will have to undergo a radical transformation, with specific programmes for this transformation to be found in Heidegger's, Simondon's and Stiegler's respective philosophical projects.

According to Heidegger, technics is the final stage of metaphysics, meaning that the historical present is in a moment of transition (*Übergang*). Our vocation is to bear witness to this transition and to prepare a different time through the act of questioning.

According to Simondon, technics is both the beginning and the end of metaphysics. It is the beginning of metaphysics because metaphysics was always already derived from the incomplete technical experience of the ancient Greek master-slave relation. And it is at the same time the end of metaphysics because, in being derived from an incomplete, socio-politically overdetermined technical experience, metaphysics is said to be structurally incapable of properly thinking technics. Simondon's own project, employing the method of problematization, aims to bring about a post-metaphysical and genetic philosophical thought, capable of thinking technics in its becoming.

According to Stiegler, the relation between philosophy and technics can best be characterized in a Derridean manner as one of repression. Stiegler's own contribution, at once Heideggerian and Simondonian, attempts to liberate technics through exhuming and thereby inventing technical thought anew. Only this way, Stiegler claims, can justice finally be done to technics.

In their respective discussions, all three thinkers have been shown to dedicate an important place to Kant. Kant himself wrote at the brink of the Industrial

Revolution and thus just prior to the widespread occurrence of industrially produced technical objects and the corresponding modern inquiry into technics. According to Heidegger, Kant's *Transcendental Analytic* laid open the a priori principles of the thingness of things, articulating the very moment in which metaphysics 'became' modern technics. Subsequently, I have argued that, according to Heidegger, the *Transcendental Analytic* could or even should have been called *Transcendental Technics*. While providing the foundations for Heidegger's own notion of enframing, Kant is, however, not read by Heidegger as a thinker of technics himself. Instead, he is read merely symptomatically, in the sense that he is said to have implicitly articulated the moment in which metaphysics became technics, without, however, self-reflexively 'understanding' it himself.

Simondon places Kant in an equally prominent position in his critique of metaphysics, which philosophy is urged to leave behind once and for all. This step is necessary for Simondon, since from within the transcendental restrictions of knowledge so forcefully articulated by transcendental philosophy, thought is either too early (a priori) or too late (a posteriori), but never contemporaneous with technical operation. According to Simondon, technics is the final question for transcendental philosophy, which it is structurally incapable of responding to; its answer will either come too early or too late. From within the history of technical thought as laid out by Simondon, Kant must also be said to have misrecognized the inherently technical nature driving the Enlightenment, which Simondon terms technical *Encyclopedism*. Subsequently then, one could argue that, from Simondon's point of view, the *Critique of Pure Reason* should have been called *Critique of Technical Reason*. While Kant's critique is directed to pure rather than technical reason, Simondon's self-proclaimed aim is to put forward a subsequent technological *Encyclopedism* to its Enlightenment predecessor, which at heart remains a humanism aimed at liberation, only this time, the liberation of technics.

Within Stiegler's project, Kant was shown to have led to the restructuration of the entire *Technics and Time* series. But similarly to Heidegger and Simondon, Stiegler also argues that Kant did not 'think' technics in its constitutive, prosthetic function, as shown, for instance, in his critical discussions of synthesis and schematization. In his engagement with the concept of technoscience, Stiegler takes further issue with the Kantian foundational distinction between theory and praxis, which, in a similar argument to Simondon, is said to be 'structurally incapable' of thinking technoscience.[1] Consequently, and despite its catalysing function within Stiegler's project, Stiegler also argues that Kant must ultimately be left behind as he marks a fundamental limit to technical thought proper.

In concluding, there are two last points to be made regarding Heidegger, Simondon, and Stiegler's double-articulation of the history of technical thought laid out in the preceding chapter. On the one hand, and first of all, the preceding chapter aimed to show, by way of Heidegger, Simondon, and Stiegler's omission of Kant's explicit technical thought and combined with their relegation of Kant to a mere symptom, how their double-articulation of the history of philosophy as at once devoid of *and* containing implicit seeds of technics, functioned somewhat as the condition of possibility for the individuation of their own technical thought. With this statement, I do not propose that their omission and symptomatic readings were done wilfully. Rather, it is my understanding that their self-positioning within the history of technical thought meant that neither Heidegger, Simondon, nor Stiegler could see what they themselves were looking for (technical thought proper in the history of philosophy prior to their own intervention). Only in this way can one understand both Heidegger, Simondon, and Stiegler's claim that they found themselves in a historically singular position from which technical thought proper could, for the first time, be developed, questioned, problematized, exhumed, invented and liberated.

Following on from this first point, the second point addresses a wider issue. What follows from Heidegger, Simondon, and Stiegler's readings of the history of philosophy as only implicitly containing the seeds for their own technical thought proper is that historical thought on technics, such as Kant's, for instance, remained veiled or in an unfortunate turn of events was veiled once again. To say with Heidegger and Stiegler that Kant does not speak of technics, 'unthought' technics, or did not ask the right question concerning technics ultimately means that Kant does not have a future beyond the status of the symptom granted to him. Heidegger, Simondon, and Stiegler's readings of Kant and their consequent development of their own technical thought out of their problematization of Kant's technical aporia would not be a problem if Kant had never written explicitly about technics proper. As the following chapters will show, however, the three technics-thinkers could themselves not see what it is they were at the same time so urgently looking for. Only in this way can the radical oversight regarding Kant's explicit writings on technics make sense. While the stand-alone engagement with Heidegger, Simondon, and Stiegler's projects has herewith come to an end, the following chapters will provide an in-depth inquiry into the role and meaning of technics within Kant's system whereby offering, at the same time, an engagement with, and critique of, the histories of technical thought told by Heidegger, Simondon, and Stiegler.

Part II

Kant's critical philosophy and technics

Any serious and fair engagement with a philosophical work must start out by articulating the problem that at once motivated the philosophical undertaking at hand and shaped the ensuing solution. When it comes to Kant, this means articulating the problem that motivated Kant to develop the critical method in order to find and justify his transcendental conditions of possibility. The point of departure is thus decided. But, one might well ask, does embarking from such a general place not risk losing the primary vantage point that motivated this inquiry in the first place? Is technics here relegated to the background before ever even seeing the light of day? As this chapter will show, it is here – at the general level of the problem addressed by Kant's philosophical project – that technics will show itself in its complex and multifaceted form for the first time. The investigation of the problem around which Kant's philosophical project orients itself thus constitutes the first part of my inquiry into the role and meaning of technics in Kant's works.

Kant discusses his critical approach in the two prefaces and introductions to his seminal work, *Critique of Pure Reason* (*Kritik der reinen Vernunft*). The double genitive in the title of the work points to the complexity of the undertaking. The uncertainty of the who and the what in the grammatical structure reveals that this is not only a critique *of* pure reason, in the sense of subjecting pure reason to critique. It is at the same time pure reason itself that carries out this critique. We are thus dealing with what could be called a self-critique, in which an internally differentiated reason is at once the court of justice, the judge and the accused themselves. But what has occasioned these extraordinary measures? What crimes has reason committed to provoke its own trial? Reason's self-critique, Kant claims, is necessary in order to settle once and for all the 'endless controversies' fought out in 'the battlefield [. . .] called metaphysics',[1] which are slowly driving the former 'queen of all the sciences'[2] into ruin. Brought about by reason's very own 'fate [*Schicksal*]', reason is said to be 'burdened with questions which it cannot dismiss, since they are given to it as problems by the nature of reason itself, but which it also cannot answer, since they transcend every [faculty] of human reason'.[3] The questions referred to by Kant are of a metaphysical order, such as: Does God exist? What is the nature of being? When did the world begin?

While these questions are given to reason's inner nature by itself, they simultaneously far exceed what reason is capable of answering. The predicament is thus indeed paradoxical, since, according to Kant, in the end reason has no one but itself to blame for its vexing dilemma. The source of the problem, Kant argues, is located in reason's nature itself. Reason is subject to a 'drive for expansion [*Trieb zur Erweiterung*]',[4] which plunges itself into ever further 'obscurity and contradictions',[5] warranting the extraordinary measures of the *Critique of Pure Reason*. It is in relation to reason's very own inner drive for expansion that technics will show itself, since, as the following discussion will show, reason's very method of expansion is instrumental in nature. In order to investigate reason's predicament in more detail, I propose the following course of inquiry. The first part of this chapter will look in more detail at reason's drive (*Trieb*). The second part will then discuss the instrumental problematic of the drive, before the third part will finally move from Kant's pathological diagnosis of reason to his twofold critical remedy, which will however itself rely, in the last instance, on a notion of technics.

5

Reason's drive

Emerging in the German language during the thirteenth century, *Trieb* first denoted a non-human driving force (*Kraft*) in the broad sense of the term, which could be applied to both technically produced objects and organic but non-human beings. From the sixteenth century onwards, the term came to refer to human beings as well, designating an internal impulse acting on the body or the mind.[1] The 1984 edition of Grimm's *Deutsches Wörterbuch* states that already during the late seventeenth century a drive (*Trieb*), generally speaking, denoted an 'internal force that impels, that puts into motion [*innere treibende Kraft*]',[2] and as such was used in relation but also differentiated from the German word *Instinkt*, which denotes an innate tendency oriented towards determinate objects or actions.[3] Within the context of the *Critique of Pure Reason*,[4] the term 'drive' (*Trieb*) appears in relation either to expansion (*Erweiterung*) or to the incentive (*Triebfeder*) of morality,[5] which the later *Critique of Pure Practical Reason* deals with explicitly. That Kant's understanding of drive was generally in line with the definition given by Grimm's *Wörterbuch* is attested to by the various references to how reason is driven by its 'own needs' (*durch ihr eigenes Bedürfnis*),[6] that it has its drive 'in itself'[7] and, finally, that it is driven by 'its own nature [*die Vernunft wird durch einen Hang ihrer Natur getrieben*]'.[8] All of these citations show that reason's drive is not a problem of external inclination, nor a mere predicate of reason. Rather, it is reason's inner nature to strive. Arising from 'within universal human reason itself [*aus der Natur der allgemeinen Menschenvernunft*]',[9] drive thus denotes the originary and continuous moving force that is reason itself.

While the drive is said to be innate to universal reason itself, the term 'drive for expansion [*Trieb zur Erweiterung*]'[10] points to its general outward orientation. What this means is that reason strives towards 'ends' or questions. And while these ends arise from 'within universal human reason itself [*aus der Natur der allgemeinen Menschenvernunft*]',[11] their answer is said to 'transcend every [faculty] of human reason'.[12] What this means, then, is that when it comes to theoretical reason as the primary object of the *Critique of Pure Reason*, it is not within theoretical reason's power to answer its own questions. This is the case because knowledge of these answers is said to lie 'outside' of theoretical reason's proper bounds. This applies in a twofold manner: reason's expansions are either aimed towards places far removed from empirical experience so as to gain access to metaphysical questions or, on the other hand, to its supposed real-world referents. Both of these ends go right to the heart of the problem of the *Critique of Pure Reason*, which as a whole sets out to decide 'about the possibility or impossibility of a metaphysics in general'.[13] Metaphysics poses questions that by definition cannot be answered by resort to experience, such as the nature of being (*metaphysica generalis*), as well as the existence and properties of God, the immortality of the soul and the beginning and end of the universe (the three domains of *metaphysica specialis*).[14] The problem articulated by Kant here is thus the following: throughout its long history, metaphysics has failed to establish itself as a science, falling into obscurities and entangling itself in contradictions and continually forcing itself to retrace its steps.[15] How, then, does reason go about achieving these ends? What is this problematic, unsound mode of expansion that Kant is talking about, by which reason is trying to answer its own questions, the answers to which at once exceed its capabilities and proper bounds?

The *organon* of reason

Kant accuses reason of employing an *instrument* or *tool* in order to achieve its expansions, for which he uses the German term *Werkzeug* (instrument), which is otherwise employed in contexts of handicraft and technical production.[16] Placing reason within a framework of instrumentality points to the fact that at stake in reason's predicament is ultimately an underlying instrumental condition of reason's drive. Reason's drive is purposive (*zweckmäßig*); consequently, it uses means (instruments) in order to achieve its ends (the questions of general and special metaphysics). But what is the ominous instrument that reason is accused of employing?

The most famous *organon* (ancient Greek for instrument or tool) in the history of philosophy is without doubt the collection of Aristotle's six logical works.[17] The name *Organon* was posthumously assigned to Aristotle's logical works by his followers and later found its way into the modern European philosophical traditions, most notably via Johann Lambert's *New Organon*[18] as well as Francis

Bacon's *Novum organum scientiarum.*[19] As pointed out by Leif Weatherby, the German term *das Organ* only gained its modern biological meaning, denoting a functional part of the organism, during the 1780s, but nevertheless retained its semantic tension between its ancient instrumental and modern biological meaning until the nineteenth century.[20] In the ancient Greek context, it was Epicurus who first critiqued Aristotle for using judgments, inferences and concepts as tools in the service of the extension of knowledge, replacing the Aristotelian *Organon* with his more humble *Canon* (ancient Greek for rule), which 'concerns criteria and first principles, or the fundamentals of philosophy'.[21] Epicurus's *Canon* aimed to first establish a body of rules for distinguishing between correct and incorrect judgments, rather than extending knowledge.[22] Kant refers to Epicurus's *Canon* in the *Jäsche Logik* and characterizes his own project as essentially Epicurean in nature.[23] Logic, according to the *Jäsche Logik*, should be nothing but a '*canonica Epicuri*',[24] comprised of the correct rules for '*passing judgments* and *correcting* our cognition, but not for *expanding* it'.[25] The *Critique of Pure Reason* as a whole proposes to restrict logic from organon to canon, whereby logic is reduced to a mere negative utility, 'namely that it does not serve for [expansion], as an organon, but rather, as a discipline, serves for the determination of boundaries, and instead of discovering truth it has only the silent merit of guarding against errors'.[26] In Epicurean fashion, Kant thus accuses reason of (ab)using logic as an instrument in its drive towards expansion.

Kant's discussion of reason's drive for expansion takes place at two prominent moments of the *Critique of Pure Reason*: first appearing during the two *Prefaces* and *Introductions*, it reappears towards the end of the *Critique of Pure Reason*, in the second of its two books. In between the *Introductions* and the second book, we find the disproportionately long *Transcendental Doctrine of Elements*, which contains the promised canon: 'a science of the mere estimation [*Beurteilung*] of pure reason, of its sources and boundaries.'[27] After reason's elements, 'sources and boundaries' have been secured by the canon, Kant moves on to the second book of the *Critique of Pure Reason*, entitled the *Transcendental Doctrine of Method*, asking what reason can do with the materials secured by the *Doctrine of Elements*. The nomenclature of the two main books, as well as their internal structuration, corresponds to common modern structurations of logical treatises with genuine Aristotelian provenance, such as the 1662 Port Royal logic.[28] What the treatment of pure reason according to elements and method points to is that reason's instrumental drive for expansion is subjected by Kant to a double treatment according to the *Critique of Pure Reason*'s two books.

Elements, sources and boundaries: The *Doctrine of Elements* and the canon

Within the context of the canon, reason's unruly expansions and the instrumental employment of logic are dealt with as a problem of proper bounds. The canon addresses the problems of bounds through a discourse on the mind's (*Gemüt*) immanent faculties (*Vermögen*) and forces (*Kräfte*) and the role that logic is assigned within that discourse. As such, it ventures to restrict what theoretical reason wants, drives or strives to know to what it 'can' know, meaning that its ends and questions are set in relation to its own proper powers, its faculties (*Vermögen*). The *Transcendental Doctrine of Elements* thus engages in the quasi-archaeological work of first analysing and then sounding the depth of the faculties (*Vermögen*) and forces (*Kräfte*) in the sense of legislating their competences. This is necessary for Kant since, as we have seen, he accuses theoretical reason of striving for ends that far exceed its faculties (*Vermögen*).[29] If reason's instrumental drive for expansion is to be averted, Kant claims that one first needs to ask, in proto-Deleuzian fashion, what reason and the faculties and forces of the *Gemüt* 'can do'.[30]

Derived from both the Latin *facultas* as well as the Greek *dynamis*, the concept of *Vermögen* (faculty) in Kant is double-coded as both the 'ability or power to achieve an end [. . . and] a potential for change which would be actualized through *energeia*'.[31] Concerning the relation between faculty and force (*Kraft*), Kant adopted the distinction from Baumgarten's *Metaphysica*, where one finds the following ontological definition: 'although positing a faculty and receptivity does not posit action or suffering, nevertheless such is possible when a power in the stricter sense is posited. This will be the complement of a faculty to act, i.e. that which is added to the faculty for the action to come to exist.'[32] This means that forces are the actualization of faculties; they must be 'added' for a faculty to be actualized.[33] The Kantian critical project as a whole thus begins by dissecting the mind (*Gemüt*) into its different faculties and forces in order to derive its adequate and corresponding ends. Consequently, reason's unruly and transcendent drive for expansion is placed within a 'restricted economy', in which reason's legitimate ends are deduced from a discussion of what it is lawfully capable of, and thus without the instrumental, but ultimately illegitimate, use of logic.

The Kantian proposal to set reason's ends in relation to its faculties can be seen clearly in his discussion of metaphysics. The preceding discussion has shown that the expansions of theoretical reason are either aimed towards places far removed from empirical cognition so as to gain access to metaphysical questions or to its supposed real-world referents. Kant's damning diagnosis of metaphysics had already started two decades prior to the *Critique of Pure Reason* in his 1763 essay *The Only Possible Argument in Support of a Demonstration of the Existence of God*. It was here that Kant first introduced the proposition that existence is not a predicate.[34] This proposition is taken up in slightly altered form again in the *Critique of Pure Reason*. Kant claims that modal judgments, which are judgments concerning the possibility, existence, or necessity of things, do not add anything to the content of any judgment. Instead, a modal judgment 'concerns only the value of the copula in relation to thinking in general'.[35] Kant thus fundamentally shifts the grounds of the metaphysical debate, proposing that, from now on, judgments need to be reflexively justified internally to thought itself.

In similar fashion to how Kant shifted the discourse of special metaphysics towards the validity of our judgments, he also ventures to displace questions about the essence of being (general metaphysics), thus effecting a radical reformulation of ontology. His transcendental idealist position proposes that 'objects of experience [. . .] are never given in themselves, but only in experience, and they do not exist at all outside it'.[36] What this means is that our concepts (as well as space and time as our forms of intuition) as a whole do not pertain to things 'out there', as a transcendental realist position would have it. But neither does Kant doubt the reality of external objects as Cartesian scepticism would. Instead, Kant promotes a more complex, twofold proposition. Taking the place of 'the proud name of an ontology',[37] Kant's *Transcendental Analytic* articulates his transcendental idealist position: while concepts and intuitions are nothing but representations, and thus have no existence outside our mind, they are at the same time the universal and necessary conditions for the possibility of things 'for us'.

Kant thus argues that one must reflexively inquire into how it is that one can judge (correctly) in order to address questions concerning both special metaphysics as well as the nature of being, and thus general metaphysics. Rather than remaining within the old metaphysical discourse with its question of 'What is X?', the Kantian project fundamentally displaces the place of inquiry. In directing our attention to the reflexive question of how we can correctly judge, Kant puts forward a fundamentally novel approach to philosophizing, which begins by investigating and securing the elements, faculties, and forces of cognition, before venturing to employ them in philosophical arguments. This undertaking of investigating, delineating, and securing the elements of cognition is the explicit subject matter of the canon, which is the first half of Kant's critical remedy to reason's instrumental drive to expansion by means of logic.

The canon and its technical critiques

In order to provide a faculty with a set of principles (*Grundsätze*) that allow it to correctly and thus lawfully go about its business, Kant argues that the canon must contain 'the sum total of the *a priori* principles of the correct use of certain cognitive faculties in general'.[38] This firstly means that a canon holds the rules for the correct use of a faculty. It secondly lays claim to totality, meaning that it must be able to enumerate all of the necessary rules of a cognitive faculty, be it the faculty of understanding, the power of judgment, or the faculty of reason. Thirdly, these rules are characterized as a priori, which means that they lie ready-for-use in the mind (*Gemüt*) rather than coming to us from experience. Fourthly, the term principle denotes their status as a 'starting point'. According to Kant, principles contain the grounds for other judgments, while they are themselves not grounded in anything higher or more universal,[39] thus implying that a principle cannot be analysed or dissolved into further constitutive parts; rather, it must be presupposed for everything else. The canon must thus hold the totality of immediately certain, a priori principles pertaining to a cognitive faculty. It is by way of the canon that Kant's critical project aims to reformulate metaphysics as the search to uncover and delineate the immanent limits of human reason, rather than abusing logic as an instrument in the search for hidden truths such as the metaphysical objects of God, freedom, and the immortality of the soul.

The correlation of a canon with the correct use of a faculty means that of our three faculties of cognition (understanding, power of judgment, and reason), only understanding and the power of judgment can have a canon in the positive sense. The canons of the faculty of understanding and the power of judgment are both expounded in the *Transcendental Analytic*. Consisting of the *Analytic of Concepts* and the *Analytic of Principles*, the two-part *Transcendental Analytic* is said to lay open 'the elements of the pure cognition of the understanding and the principles without which no object can be thought'.[40] However, the canon constituted by reason, as our third faculty of cognition, consists in a *Transcendental Dialectic*, rather than an *Analytic*. Here, the canon functions as a 'critique of the understanding and reason in regard to their hyperphysical use, in order to uncover the false illusion of their groundless pretensions'.[41] The *Dialectic* thus lays open the logic of illusion in reason's misguided attempts to syllogistically expand our knowledge of the special metaphysical objects of the soul, freedom, and God. This means that, differently to the theoretical faculties of understanding and judgment, reason in its theoretical use only has a negative canon, which illustrates the contradictions of reason if it fails to proceed according to its critical limits. At the same time, reason nonetheless acquires a positive use of its canon once the discussion moves beyond theoretical reason. But since it will then be a canon for reason in its practical rather than theoretical use, it is only the second *Critique of Pure Practical Reason*, that explicitly expands on it.[42] As such, this positive canon will not feature in the present discussion which focuses on the *Critique of Pure Reason*.

Within the domain of the *Critique of Pure Reason*, Kant consequently specifies two positive canons: a canon for the faculty of understanding and a canon for the power of judgment. Concerning the canon of the faculty of understanding, Kant proposes that general logic is nothing but the pure formal law of the understanding. In laying out the rules for thinking so that thought is in accord with itself, general logic contains the mere formal rules of thought, and, as such, it is to be sought and tried internally to thought itself. As a canon for the merely formal use of the understanding, general logic, however, does not suffice to show and secure how the use of the cognitive faculty by itself is possible. This is the case because the canon of general logic is not only devoid of content but, furthermore, completely indifferent to any content provided to it from elsewhere. Thus, if there is to be a 'lawful' relation between reason's forms and any possible content of thought (be it the actual world of objects or metaphysics), which does not rely on the transgressive use of logic as an instrument, there needs to be something else, something 'more', to make possible this very relation between theoretical reason and its other.

And indeed, Kant proposes that complementing the first, purely formal, outline of the canon of general logic, there is another side to the canon of the faculty of understanding, which comes onto the scene in order to secure precisely this sought-after non-instrumental relation of reason's forms to its content. Going by the name of *Transcendental Logic*, Kant argues that it is 'a logic in which one did not abstract from all content of cognition'.[43] Kant's claim is nothing less than that transcendental logic is at once pure and endowed with content at the same time. How is such a logic possible?

Famously distinguishing between the understanding and intuition as the two sources of cognition, Kant's transcendental idealism proposes that 'in whatever way and through whatever means a cognition may relate to objects, that through which it relates immediately to them, and at which all thought as a means is directed as an end, is intuition'.[44] On the one hand, there is the understanding with its universal concepts for thinking objects, and, on the other hand, there is sensible intuition (*Anschauung*), which is charged with relating these concepts to actual objects. This means that Kant's critical remedy to the illegitimate (ab)use of logic as an instrument hinges on his claim that objects can only be 'given' to us in the sense of being affected by them in sensibility.[45]

Of our two faculties of cognition, sensibility (*Sinnlichkeit*) is thus identified by Kant as the faculty through which objects can be given to us and through which cognition may relate to objects. However, the clause 'in whatever way and through whatever means cognition may relate to objects' indicates that there are different ways in which cognition can relate to objects. In fact, intuition is itself said to have two sides. Differentiating between the form and the matter of intuition, Kant argues that while the matter of intuition is always subject to being empirically given, the same does not apply to the forms of intuition. Constituting a curiously non-sensible side of sensibility, the forms of intuition are specified as nothing other than space and time, which, Kant claims, lie a priori ready in our mind (*Gemüt*) in a similar way to the concepts of the understanding. Thus, while the matter of intuition is always dependent on a particular experience of a given object, the forms of intuition can never be given experience.

The receptive character of sensibility is consequently complemented by the forms of intuition, which, in a similar manner to the concepts of the understanding, can be anticipated and enumerated prior to any particular experience.[46] Specified as space and time, the a priori forms of intuition are what

> allows the manifold of appearance to be intuited as ordered in certain relations [. . .]. Since that within which the sensations can alone be ordered and placed in a certain form cannot itself be in turn sensation, the matter of all appearance is only given to us *a posteriori*, but its form must all lie ready for it in the mind *a priori*.[47]

It is intuition's non-sensible form, which is explicitly said to allow for the given content of intuition to be 'ordered'. The forms of intuition are thus charged by Kant with the action of ordering (*ordnen*) and placing (*stellen*) the matter of intuition given to us.[48] Thus, if an object is to be received by us, it is explicitly subject to being placed and ordered according to the a priori forms of intuition.

Lingering for a moment on the word placing (*stellen*), it appears that the preceding discussion of the a priori forms of intuition, which are specified by Kant as what allows reason to relate to things outside of itself rather than illegitimately extending itself by means of logic alone, might have led us back to what Heidegger calls enframing (*Gestell*). In *the Question Concerning Technics*, Heidegger claimed that the essence of modern technics as enframing (*Gestell*) is at once an ordering attitude (*bestellendes Verhalten*) as well as a quasi-'hunter' that 'pursues and entraps [*nachstellen*] nature'[49] to reveal itself as a standing reserve (*Bestand*). This ordering attitude of enframing (*Gestell*) was then argued to have first shown itself in modern metaphysics and specifically in Kant's *Critique of Pure Reason*. For Heidegger, the placing and ordering (*stellen*) of the matter of intuition into the forms of space and time thus do not denote neutral or even harmonious acts that make the reception of sense objects possible. Instead, they are the transcendental condition for the technical domination of nature. The acts of placing (*stellen*) and ordering (*ordnen*) performed by the pure forms of intuition are, Heidegger argues, fundamentally a technics of entrapping and pursuing (*nachstellen*).[50] The verbs *stellen, nachstellen, bestellen* all denote the operations by which enframing (*Gestell*), as modern metaphysics' heir to the mind (*Gemüt*), challenges the real to reveal itself as an identifiable and spatio-locally determinable number of scientific objects.

Heidegger's reading goes right to the heart of the Kantian problematic. Kant had put forward the canon in an attempt to restrict reason's instrumental drive towards expansion by means of logic. And it was to this end that the canon charged intuition, rather than the logic of the understanding, with relating to its 'other'. Heidegger's point is thus that, in critically avoiding the (ab)use of logic as an instrument and by charging intuition rather than logic with the responsibility for relating to the world of things, Kant came to articulate, in a quasi-ironic fashion, nothing less than the very essence of modern technics as enframing (*Gestell*), an insight which is however said to have escaped Kant himself. Consequently, the preceding chapter has argued that if one were to think Heidegger's critique of Kant in relation to the question of enframing to its conclusion, the *Transcendental Analytic*, by which Kant aimed to replace 'the proud name of ontology',[51] should have properly been called a *Transcendental Technics*.

Heidegger was, however, not the only one to put forward a technical critique of Kant's canon. Differently from Heidegger, Stiegler does not identify the essence of modern technics with Kant's remedy to reason's instrumentality but instead argues that Kant's pure and a priori canon is made possible by a 'subterranean' technics. Against Kant's claim of treating reason in general, Stiegler then insists that it is not possible to consider the faculty of pure reason in general without its material, technical conditions and its embodiment in 'books and systems'.[52] For Stiegler, Kant's insistence on a pure faculty of reason did not get rid of its technical conditions but demoted them to a subterranean status all the while relying on them.[53] Heidegger and Stiegler thus read Kant's canon, explicitly put forward in order to restrict reason's instrumental drive for expansion, as either articulating nothing less than the essence of modern technics or as relying on a repressed, subterranean technics.

Interestingly enough, Heidegger and Stiegler were not the first to critique Kant's pure canon in technical terms. As a matter of fact, this very accusation was already directed at Kant during his lifetime. And as the following will show, Kant furthermore responded to this critique. It thus appears that, in their insistence on the implicit status of a technical problematic in the canon, both Heidegger and Stiegler failed to account for a discussion that Kant himself partook in, which centred precisely on the question of the technical problematic of the canon.

The canon, an a priori manufacture?

As pointed out by Catherine Malabou in *Before Tomorrow: Epigenesis and Rationality*,[54] the German Enlightenment thinker and cousin of J. W. von Goethe, J. G. Schlosser, critiqued Kant in his 1795 work *Plato*'s *Briefe nebst einer historischen Einleitung und Anmerkungen*. Schlosser's accusation goes right to the heart of the preceding discussion, as he accuses Kant of having reduced cognition to 'a mere form-giving manufacture [*Formgebungs-Manufactur*]'.[55] The term *Manufactur* (Latin: *manus* hand; *facere* to make and produce) refers to the historical moment in between artisanal and industrial production, in which previously decentralized crafts were first united under one roof and according to a common aim, as, for instance, in the gathering of turners, metal workers and goldsmiths under the roof of a carriage manufacture. Schlosser's critique contests that Kant's critical philosophy has abstracted from all matter (*Materie*) of cognition. As a pure system of forms, it presents itself as a mere problem of shaping – of giving form to matter by way of the interplay of the faculties, just like a manufacture. Schlosser thus accused Kant during his lifetime of having lost all content of cognition and of having transformed philosophy and in specific the a priori into a proto-industrial form-giving mechanism that shapes the world in its image.

Kant responded to Schlosser's accusation one year later in his 1796 essay *On a Recently Prominent Tone of Superiority in Philosophy*. He referred to Schlosser's critique as 'the dismissive habit of crying down the *formal* in our knowledge (which is yet the preeminent business of philosophy) as a pedantry, under the name of "a form-giving-manufactur" [*Formgebungsmanufaktur*]'.[56] Distancing the specific kind of 'diligent and careful work of the subject [*fleißige und sorgsame Arbeit des Subjects*]'[57] performed by the cognitive faculties from the accusation thrown at him by Schlosser, Kant explicitly writes that the work of this subject is neither an 'arbitrary form-giving undertaken by a plan or even factory-like [fabrikenmäßig] (on behalf of the state)'.[58] While the translators of the Cambridge University Press translation have translated *fabrikenmäßig* as 'machine-made', the original German term refers not to the machine but to the factory. In the eighteenth and nineteenth centuries, the terms 'manufacture' (*Manufactur*), 'workshop' (*Werkstatt*) and 'factory' (*Fabrik*) were often used synonymously,[59] and the qualification '(on behalf of the state)'[60] refers to the French model of King Henri IV, by which manufactures were initially state-founded, a practice that was soon adopted throughout Europe.[61] Kant distinctly distances himself from Schlosser's accusations, introducing a difference between the preceding 'diligent and careful work of the subject [*fleißige und sorgsame Arbeit des Subjects*] to takeup and evaluate its own faculty (of reason)',[62] on the one hand, and the 'manufacture of handling the given object',[63] on the other. The work of the subject does not proceed according to an arbitrary, pre-existing pattern or design, provided by others or even one's own work; there is no pre-planned and anticipated design at work in this a priori 'working' on transcendental forms.

A priori operations (*Handgriffe*)

The exchange between Kant and Schlosser is indeed remarkable, as it includes Kant's response to Schlosser's critique of the canon. As the above has shown, Kant's response essentially argues for an autonomous and spontaneous a priori work of the subject in contradistinction to the manufacture of dealing with empirically given objects. In *Before Tomorrow: Epigenesis and Reason*, Catherine Malabou problematizes Kant's response due to reintroducing 'the risk of innatism. How do we defend the idea of a "pure labor" without assimilating it, quite simply, to a lack of labor, to a mystery, to a gift, once more?'[64] This work, which is said to be the responsibility of the autonomous subject and its reason, can refer to nothing other than the work of synthesis and schematization. In the *Critique of Pure Reason* Kant explains that synthesis, in the most general sense, is 'the action of putting different representations together with each other and comprehending their manifoldness in one cognition'.[65] From the ancient Greek *syn*, 'together', and *tithenai*, 'to put, to place', synthesis is thus the name for the act of assembling and unifying different representations into one. What fundamentally differentiates synthesis from its counterpart analysis is that, while analysis aims at clarifying and taking apart something that is given in its entirety, synthesis is always excessive and thus expansive. This means that something can, and, in fact, must always be added to a given concept for a judgment to be synthetic rather than analytic. Synthesis thus adds something to the concept, something that is true, but this truth cannot be perceived internally to the concept itself according to the formal principle of non-contradiction. The problem of synthesis, then, is how to lawfully step outside of the concept, or, to put this differently and in reference to the aim of the canon as restricting the (ab)use of logic as an instrument, how to lawfully expand a concept by means other than logic.

In its more specific form, synthesis designates the act of schematization, which is the procedure of putting forward, in the sense of synthesizing, the famous 'thirds', by which a category of the understanding is supplied with a sense impression. These thirds are necessary since the *Critique of Pure Reason* famously sets off, as we have seen, from the fateful proposition that there are 'two stems [*Stämme*] of human cognition, which may perhaps arise from a common but for us unknown root [*Wurzel*], namely sensibility and understanding'.[66] And while objects can only be given to us in sensibility, the contribution of the faculty of understanding ensures for these given objects to be thought. From hereon, then, due to this essentially twofold nature, cognition is in need of unification, the search for and supplementation of which will occupy Kant for the remainder of his intellectual life (and most if not all post-Kantians after him). Within the constraints of the first book of the *Doctrine of Elements* of the *Critique of Pure Reason*, Kant charges the power of imagination (*Einbildungskraft*) with the necessary and vital work of weaving together the two disparate stems. What emerges from the power of imagination's efforts of unification are a number of in between 'thirds'[67] – the famous schemata – which are charged with connecting concepts, the representations supplied by the faculty of understanding, and intuition, the representations supplied by the faculty of sensibility. As products of the power of imagination, schemata are, however, not mere images and neither are they examples or symbols. Rather, Kant explains that they provide our power of judgment, the faculty responsible for applying concepts to phenomena, 'with a rule under which given empirical intuitions are to be subsumed'.[68] Subsequently, only by way of the schematism does it become possible for a category to apply to, in the sense of lawfully connect to, the sensible world. Only by way of these thirds are our concepts supplied with content, whereby they gain sense and meaning (*Sinn und Bedeutung*).[69] Only by way of these thirds is knowledge proper possible for the human subject, ensuring that our categories have objective, rather than merely subjective validity. Only by way of these thirds is it that we are not merely playing with words, but that these concepts correspond to and are capable of calling upon a world. When pressed to lay out the precise workings of schematization, however, Kant admits that 'this schematism of our understanding with regard to appearances and their mere form is a hidden art in the depths of the human soul, whose true operations we can divine from nature and lay unveiled before our eyes only with difficulty'.[70] Kant thus concedes that it is impossible to lay open exactly how schematization in fact takes place. It is, he says, a 'hidden art [*verborgene Kunst*]'.[71]

Essentially recoiling from the demand of not only description but explication, Kant characterizes schematization as a 'hidden art', whereby introducing a notion of skill and ability into the heart of the a priori workings of the subject, that is, the transcendental domain. This hidden art of schematization refers to theoretical reason's skill of bringing forth these 'thirds' which mediate between concepts and intuitions and which belong neither to sensibility nor the understanding and are thus neither exclusively sensual nor theoretical.

It is remarkable to see Kant take recourse to the term art (*Kunst*) here, since it is mostly used in the *Critique of Pure Reason* to refer to the undesired organon and in reference to sophism. If the canon of general logic is '[mis]used as if it were an organon for the actual production [*Hervorbringung*] of at least the semblance of objective assertions',[72] it is always a 'logic of illusion – a sophistical art'.[73] There are, however, four different mentions of the word 'art' in the *Critique of Pure Reason* that deviate from these negative uses. The first positive use of the term refers to the restriction of philosophy's function to critique, employing philosophy's 'perspecacity and art of scrutiny [*Prüfungskunst*]'.[74] The second positive use of the term refers to the characterization of schematization as a hidden art.[75] The third mention of art outside the context of sophism can be found in reference to the human art (*menschliche Kunst*) of production (*Hervorbringung*).[76] The last noteworthy use of the word art can be found in the third chapter of the *Doctrine of Method*, entitled *The Architectonic of Pure Reason*. The opening sentence of this section reads as follows: 'By an architectonic I understand the art of systems [*Kunst der Systeme*]. Since systematic unity is that which first makes ordinary cognition into science, i.e., makes a system out of a mere aggregate of it, architectonic is the doctrine of that which is scientific in our cognition in general [. . .]'[77] In a paradoxical manner, then, the general notion of art (*Kunst*) in the *Critique of Pure Reason* refers at once to the organon and sophism, schematization, human art in general, and the 'art of systems', referring to reason's natural systematicity.

The notion of art employed in the schematism chapter belongs to a general notion of art that takes its cues from *Können* (ability, skill).[78] As ability or skill, this notion of art gains its definition in relation to Aristotle's *techne* (art, technics), which distinguished the art of production (*poiesis*) from theoretical knowledge and science (*episteme*).[79] Kant's *Critique of the Power of Judgment* (1790) will continue along this Aristotelian outline, when Kant argues that ability and skill (*Können) are* indeed to be differentiated from knowledge (*Wissen*). Art as skill (*Geschicklichkeit*) essentially denotes a practical faculty distinguished from a theoretical one, as technique [*Technik*] is distinguished from theory (as the art of surveying is distinguished from geometry). And thus that which one can do as soon as one knows what should be done is not exactly called art. Only that which one does not immediately have the skill to do even if one knows it completely belongs to that extent to art.[80]

Thus, in the *Critique of the Power of Judgment*, art (*Kunst*) that is dependent on skill (*Können*) is part of the practical domain of *Technik* rather than that of theory or science. But how are we to make sense of this reference to art, technics and thus the practical domain within the context of the *Critique of Pure Reason* and the question of the canon of pure (theoretical) reason, that is, the 'science of the mere estimation [*Beurteilung*] of pure reason, of its sources and boundaries'?[81]

The *schematism* chapter is located in the second part of the *Transcendental Analytic* and as such within the first book of the *Critique of Pure Reason* entitled *Doctrine of Elements.* Its precise place is, however, in the first section of the twofold *Transcendental Analytic of Principles.* Putting forward a canon for the power of judgment, the *Analytic of Principles* does not ask about the condition of possibility of judgment and thus how judgment is possible, but rather asks how judgment, concerning the subsumption of particulars under universals, can be done correctly. In surprising fashion, then, differently to the preceding *Analytic of Concepts*, the *Transcendental Analytic of Principles* does not ask a transcendental question of possibility *per se*. In asking how to correctly judge and give concepts to intuitions and vice versa, it appears to be closer to the problematic at the heart of the *Transcendental Doctrine of Method*, which asks about what we can do with the elements first secured by the *Doctrine of Elements*. The second book of the *Critique of Pure Reason*, while nevertheless being a part of the first *Transcendental Doctrine of Elements*, thus essentially consists of a methodological inquiry into judgment and introduces a notion of art as skill and ability into the heart of the canon. According to Kant, there is thus an art and technics proper to the transcendental domain, that is, the hidden art (*Können*) of schematization. In determining schematization as hidden, however, Kant forecloses the possibility of true insight into the art of this transcendental art of judgment, this art of subsuming particulars under universals. Thus, in a certain sense, Heidegger and Stiegler are right in their assessment of the implicit yet constitutive role played in the canon by technics. The explicit designation of the schematism, upon which all knowledge depends in the last instance, as a hidden art and thus as not subject to insight means that 'the *Critique of Pure Reason* is founded upon an aporia', as Howard Caygill writes in the *Art of Judgment*.[82]

What we do know, however, is that schematization is nevertheless explicitly called an art, be it hidden or not. And this hidden art is subject to one more important determination of its functioning. Schematization, Kant writes, takes place by way of a number of '*wahre Handgriffe*'.[83] The translators of the Cambridge translation of the *Critique of Pure Reason* have translated the German '*wahre Handgriffe*' as 'true operations'.[84] However, while the English term 'operation' derives from the Latin *opus* as a work or product of the general notion of art,[85] it lacks reference to the hand, which is one of the constitutive stems of the composite German term employed by Kant. What is further lost in the English translation is Kant's reference to the action of grasping (*greifen*), common to both the German term *Begriff* (concept), which, as a discursive form of unity, 'grasps' and unifies the matter of presentation, and *Handgriff* (operation), which designates the action of grasping or seizing something by hand. While in German both nouns explicitly refer to the act of grasping, the characterization of the act of combination as a *Handgriff* transfers the problematic of unification from the domain of discursivity and conceptuality to that of the hand and technical-manual skill. The unifying procedure of a *Handgriff* designates both the act of grasping by hand and the form and unity of a grasping hand, both of which retain their relation to the skill of working by hand. By determining the art of schematization as proceeding by way of *Handgriffe* rather than *Begriffe*, Kant thus seems to propose that schematization is not guided in its act of combination by yet another discursive rule and so on ad infinitum. By transferring the rule of unity from the domain of the concept (*Begriff*) to that of the operation (*Handgriff*), Kant aims to avoid the aporia of judgment, at least in theory.

Despite Kant's objection to calling his sought-after a priori synthesis a *Manufaktur*, schematization thus turns out to be a literal manufacture: the art of schematization is specified as the hidden art of grasping by hand. But rather than referring to the economic reality of a pre-industrial workshop or factory, which unified several mechanical arts under one roof and according to one aim, the manufacture performed by the art of schematization is said to be a free and spontaneous production by the power of imagination and as such distinguished from processes of mechanical production according to pre-existent design. As a number of *Handgriffe*, it is said to take place without the rule of prior conceptual forms of unity provided by the faculty of understanding, which is the very point that Kant's 1796 response to Schlosser essentially rests on. The difference between the act of grasping by hand (*Handgriff*) and the discursive unification by means of concepts (*Begriff*) is meant to make sure that schematization indeed takes place freely and spontaneously, and thus without recourse to ever further conceptual, discursive rules provided by the faculty of understanding. In the last instance, Kant's canon of transcendental logic subsequently derives its self-proclaimed superiority over general logic by means of the art of schematization and its set of technical-manual *Handgriffe*, which ultimately provide the ground for judgment and the secure knowledge claims that flow from it. In the last instance, the canon thus relies on an art (*Können*).

6

Instrumental expansion and the Transcendental Doctrine of Method

The preceding section argued that the first route taken by Kant in his critical attempt at restricting reason's (ab)use of logic as an instrument by means of a canon has problematically led to the art of schematization, understood as the skill of grasping by hand. While I have argued that this art is essentially a technical problematic, within the *Doctrine of Elements* it is, however, not explicitly called technics (*Technik*), but instead referred to under the more general term of art (*Kunst*).

However, the canon was only one of the two ways by which Kant aimed to critically restrict reason's instrumental drive for expansion. The second route of critical restriction laid out by Kant in the *Critique of Pure Reason* takes place in the second book, entitled *Transcendental Doctrine of Method*. In distinction from the canon put forward in the *Doctrine of Elements*, which took inventory of all of reason's rightful materials and elements (its faculties and forces), the *Transcendental Doctrine of Method* follows up by asking what reason can do with these materials. Engaging the affordances of reason's materials through the metaphor of building, it is 'concerned not so much with the materials as with the plan, [. . .], we have to aim at an edifice in relation to the supplies given to us at the same time suited to our needs'.[1] This means that the *Transcendental Doctrine of Method* engages the sum total of reason's secure and a priori elements laid out by the canon as if it were a building and sketches its systematic outline. It thus functions as somewhat of a manual for building, addressing both the question of what kind of house reason wants to build as well as what it can realistically afford to build.

Kant's reference to building, sketching plans and bringing forth systems is by no means a mere rhetorical device. In the *Jäsche Logik*, one finds references to method as a 'logical art [*logische Kunst*]',[2] which is said to properly belong to the 'technical'[3] part of logic. Thus, while the canon as the first half of Kant's critical remedy to reason's instrumentality was already shown to rely on the art (*Können*) of schematization as the technical-practical skill of grasping by hand (*Handgriffe*), Kant explicitly terms the subsequent method, which contains the second part of Kant's critical restriction of reason's instrumental drive for expansion, technical. As I will argue in the following, the *Doctrine of Method*, while itself explicitly aimed at restricting reason's instrumental drive of expansion, is further proof of Kant's explicit engagement with technics (*die Technik*). As I will argue in the following, Kant developed his concept of technics in the context of the *Critique of Pure Reason* out of two parallel genealogies.

The emergence of the German terms *die Technik* and *die Technologie*

In order to understand Kant's employment of the adjective 'technical' (*technisch*) in the above citation, I propose to engage with the work of German linguist Wilfried Seibicke, who in 1968 published a study entitled *Technik – Versuch einer Geschichte der Wortfamilie um τέχνη in Deutschland vom 16. Jh. bis etwa 1830.*[4] Seibicke's historical linguistic study traces the German terms *die Technik* and *die Technologie* from their first emergence in the sixteenth century until the first half of the nineteenth century, where their usage became so widespread and frequent that a genealogical linguistic study becomes impossible. This philological excursion is important as nothing proves a priori that the German terms *Technik* (technics) and *technisch* (technical) meant the same during Kant's time as they do today. How, then, did the German terms *die Technik* (technics) and *die Technologie* (technology) develop within the German language, and what role, if any, did Kant play in the development of these terms?

By means of an introduction, Seibicke sums up the state of literature in four comprehensive and paradoxical points, which have partially also surfaced in the preceding discussions of Heidegger, Simondon, and Stiegler's accounts of the history of technical thought. Seibicke argues that, first, the German word *die Technik* is often considered to stand in direct relation to the ancient Greek term *techne*, which would function as either model or counter-model for the modern German term. Second, accounts leap from the ancient Greek concept directly to the Industrial Revolution, which is said to be responsible for having instigated the development of the modern concept of *die Technik*. Third, the modern concept of *die Technik* is either treated as a completely new phenomenon or as a mere stage of the development of *homo technicus*.[5] And fourth, lexicographical sources claim that the German terms have been adopted from French and 'germanized' during the eighteenth century.[6] The aim of Seibicke's project is to systematically refute all four of these points via a discussion of German texts and dictionaries from the first emergence of the word root *techn-* in the scholarly Neo-Latin of the sixteenth century up until the first half of the nineteenth century.

With regards to the first and fourth points, he claims that there is no sufficient evidence that the German terms have been adopted and developed from the French language. Concerning the modern German word *die Technologie*, Seibicke shows that the term was in fact first introduced in writing in Germany in 1777 by the German scholar Johann Beckmann, a contemporary of Kant's. What is important to note about Beckmann's intervention is that he explicitly took recourse to the Greek word root as mere 'linguistic matter', without, however, seeking to transpose the historical meaning of the term.[7] Rather, Beckmann's aim was to employ the term *die Technologie* for his newly invented science of the comprehensive, systematic treatment of handicrafts and industrial processes, considered from the point of view of transforming natural resources into products.[8] Due to the popularity of Beckmann's lectures at the University of Göttingen, later published in his book *Anleitung zur Technologie* (1777), his scientific reputation grew rapidly throughout Europe, and it was not long until the new meaning of the German term *Technologie* was subsequently adopted into both French and English. Seibecke's argument is supported by ample lexicographical evidence, which shows that the respective modern French and English meanings of the term in fact only begin to appear in French and English dictionaries from 1800 onwards, thus following Beckmann's new science of *Technologie*.

The French philosopher of science and technology Jan Sebestik grants an equally important position to Beckmann, as he contextualizes Beckmann's efforts of establishing the science of *Technologie* in Germany in relation to developments in both France and England. His claim is that eighteenth-century Germany felt a necessity 'to fill the scientific and the technological gap'[9] after the long wars of the seventeenth and eighteenth century. Next to the quickly industrializing Great Britain and the emerging science of *Economy*, developed by Adam Smith and Ricardo, as well as the artisanal and politically advanced France with its *Encyclopedie* by Diderot and D'Alembert, Germany was both economically ruined and politically backwards. Filling in that gap, Beckmann's science of *Technologie* was, however, not meant for the wider population or artisans themselves like the *Encyclopedie*. Rather than offering detailed descriptions or dealing with issues of technical pedagogy, the science of *Technologie* was initially intended as a strategic, specialized discipline, and developed by Beckmann for both politicians and state managers, many of whom at one point or another passed through his teaching.[10]

Importantly, Kant did not refer to Beckmann's empirical science of *Technologie* when he called methodology the technical part of logic. Seibicke shows how, parallel to the eighteenth-century German development of the science of *Technologie* by Beckmann, the term had already lived a long life in an altogether different domain. In fact, the Latin term *technologia* and a whole host of derivative terms had been employed since the sixteenth century in scholarly logical and dialectical debates. The philosophical term *technologia* first appeared in France in 1565, in *De methodo libellus* by Adrianus Turnebus, before the German protestant metaphysician Clemens Timpler employed it in his 1606 *Metaphysica systema methodicum*. The scholarly meaning of *technologia* emerged out of the confrontation between the Aristotelian definition of *techne*, understood as a subjective habit (*hexis*) and the Stoic definition of *techne* as a system of rules.[11] With the Stoic definition taking the upper hand, *technologia* denoted the systematic organization of the disciplines of the *ars liberalis* from then on.[12] Heinrich Alsted adopted the term *technologia* from Timpler in his 1610 work *Panacea philosophica, id est facilis, nova et accurate methodus docendi et discendi universam encyclopedia*, with one section entitled *De Canonica et Technologia*. Two years later, Alsted differentiated between four *praecognita* in the general theory of knowledge put forward in his *Philosophia digne restitua*. The first *praecognita* was *Archeologia* concerned with eternal truths, followed by *Hexiologia* as the doctrine of habitus and the different kinds of knowledges. The third *praecognita* was called *Technologia*, dealing with the systematic properties of the sciences in relation to one another, with the final *Canonica* containing the pedagogic doctrine of rules.[13] In seventeenth-century philosophical scholasticism, the term *Technologia* thus designated the systematization of the sciences and the arts. This is supported by an entry in the 1653 philosophical dictionary *Lexicon philosophicum* by Joh. Micraelius, which explicates *Technologia* as '[. . .] the doctrine of the properties, about the order and the number of disciplines, that are either philological, as grammar, rhetorics, & c. or philosophical, both theoretical as practical; or the higher faculties, as theology, Medical & Juridical'.[14]

Apart from denoting the systematization and doctrine of the arts, from the seventeenth century, the philosophical notion of *technologia* also began to designate the terminology specific to an art, usually referred to as *terminus technologicus*. Johann Walch's entry on the newly Germanized term *Technologie* in his 1726 *Philosophisches Wörterbuch* reads as follows:

> Technology [*Technologie*] is the name of the doctrine of technical terms [*Kunst-Wörtern*], which generally designates words that refer to things that are peculiar to people in a certain trade. [. . .] They can be divided into unlearned and learned parts, the first of which refer to the handicrafts and common arts; the latter refer to the learned crafts and common arts; these occur in the learned sciences and are either general or special; The general ones that are used by all scholars are of philosophy, especially in the so-called ontology, e.g. causa, effectus, principium, principatum, subiectum, adiunctum, substantia, accidens and the like, but the special ones in theology, legal scholarship and medicine.[15]

This means that from the beginning of the eighteenth century onwards, the German term *Technologie* also referred to the doctrine of technical terms specific to a discipline. And, as such, it applies not only to the special sciences and arts but also to philosophy, with ontology providing the most universal technology for all other sciences.

Seibicke shows that it was in the work of German philosopher Christian Wolff, a contemporary of Walch, that the German term *Technologie* was first applied to the artisanal crafts, thus providing an important step towards the general, contemporary meaning of *die Technik*.[16] Wolff's aim was to build a philosophical system that would include all human knowledge and rational action. Consequently, this system was to include the mechanical-practical arts. Thus, the Wolffian account of *Technologie* must not be understood as a doctrine of art words or technical terms (*Kunstwörterlehre*) specific to a science or trade, as was still the case for Walch. Instead, it is explicitly called a *Kunstlehre* (doctrine of art), which concerns the practical, artisanal and mechanical arts themselves.[17]

Following Wolff, Seibicke then moves on to credit Kant with, first, shortening the medieval scholastic notion of *Technologie* into *die Technik* and, second, introducing the shortened term into German philosophy proper. In between the years 1781 and 1800, thus excluding Kant's pre-critical as well as post-critical works, Seibicke finds a total of seventy-one explicit references to *die Technik, technisch* and *technicism* in Kant's published writings.[18] Due to the frequency of use, Seibicke concludes that these terms must have belonged to Kant's steady vocabulary, even though, in certain instances, Kant did not always follow general linguistic usage.[19] Seibecke roughly identifies two complementary meanings of the term *Technik* in Kant's works. On the one hand, he shows that Kant employs the term *Technik* to refer to the method of systematization and the doctrine of technical terms. On the other hand, Kant also reintroduces the Aristotelian notion of *techne* as skill and ability, which we will turn to in the following chapter.[20] For the present discussion, it suffices to emphasize that the first meaning of technics in Kant refers directly to the old scholastic *Technologie* and thus to the problem of technical terminology and systematization. One can see this meaning of *Technik* at play, for instance, when Kant writes in the *Critique of Pure Reason* that his manner of division and systematization of philosophical categories surely differs from that of the '[technics] of the logicians'.[21] It is also employed in this sense in the previously discussed *Jäsche Logik*, where Kant writes that '[. . .] logic would thus have a *dogmatic* and a *technical* part. The first would be called the *Doctrine of Elements*, the second the *Doctrine of Method*. The practical or technical part of logic would be a logical art in regard to order and to logical terms of art and logical distinctions'.[22] Both of these references show that Kant employed the notion of *Technologie*, now shortened to *die Technik* and *technisch* to refer to the systematization of technical terminology intrinsic to questions of methodology. The entire second book of the *Critique of Pure Reason* thus essentially contains the *technical* part of the *Critique*. Seibecke proposes that Kant's shortening of the term *Technologie* into *Technik* might have been due to the rapid adoption of the new meaning of *Technologie* after Beckmann's 1777 *Anleitung zur Technologie*. Kant might have decided to shorten the term used to designate the scholarly concept of systemization of concepts and rules to *Technik* in an attempt to differentiate the older, philosophical meaning of *Technologie* from Beckmann's newly founded science.[23] The employment throughout Kant's lectures on logic of the Latin *termini technici*, designating the technical terms specific to a science and always in reference to problems of method, gives further indication of Kant's employment of and engagement with the medieval notion of *Technologie*.[24]

Seibicke's thus shows, by way of his linguistic study, that within the German university of the eighteenth century, there were two complementary strands of thought concerning the terms technics (*Technik*) and technology (*Technologie*). On the one hand, Beckmann invented the new science of technology with his 1777 book *Anleitung zur Technologie*. On the other hand, Kant shortened and employed the term *Technik* in contradistinction to Beckmann, but in direct relation to the medieval scholastic notion of *Technologie* as a problem of philosophical methodology. In a surprising turn of events, Seibicke's inquiry into the history of the German word *die Technik* thus revealed Kant as having shortened *Technologie* into *Technik*, as well as successfully introducing the shortened term into German philosophy. Kant's frequent use of the Germanized and shortened word furthermore suggests, according to Seibicke, that the term was quickly adopted both by Kant's contemporaries and later readers, which can be well observed in the adoption of the Kantian terminology in subsequently published works such as those by Schiller or Goethe. Importantly, Kant's use of the term *die Technik* in reference to problems of method derives its meaning from the scholastic notion of *techne*, and thus from an idea of technics that is not intrinsically differentiated from the systematicity of science. Rather, technics here designates a problem of theoretical philosophy, referring to the methodological problem of systematization and terminology. Essentially, then, the terms 'technics' (*Technik*) and 'technical' (*technisch*) in Kant are at root not a problem of instrumentality, use or even an intrinsically empirical phenomenon, already indicating that his explicit notion of technics sits uneasy within the narrative of metaphysical technical thought told by Heidegger, Simondon, and Stiegler, where technics since ancient Greece has always been employed in contradistinction to the systematicity of science so forcefully articulated by Aristotle. Paying attention to the particular genealogy of the German term *die Technik* essentially troubles the simple, narrow account of the history of technical thought told by Heidegger, Simondon, and Stiegler. Seibicke's prominent inclusion of Kant in the development of the German term *die Technik* challenges Heidegger, Simondon, and Stiegler's assertions that Kant could not, or simply did not, think technics; the preceding excursion into Seibicke's study has clearly shown that Kant thought technics. At the same time, however, technics meant something different for Kant than it did for Heidegger, Simondon, and Stiegler, the details and history of which has hitherto escaped both Heidegger, Simondon, and Stiegler. The remainder of this book is dedicated to its explication.

Technics in the Transcendental Doctrine of Method

Returning to the technical *Transcendental Doctrine of Method* in light of the preceding discussion of Seibicke's study, one can see that Kant's method – the technical part of logic – is organized according to four parts: the discipline, the canon, the architectonic and finally the history of pure reason in its theoretical use.[25] The first chapter on discipline directly addresses reason's mode of expansion, meaning its method of 'building', while the canon lays out the rightful extent of what can be built. The third chapter on the architectonic then deals with the form of unity employed in the process of building, while the fourth and final chapter concludes the *Critique of Pure Reason* with Kant's brief history of the chief revolutions of theoretical reason according to the concept of the object, modes of knowledge and in respect to method.

As the preceding discussion on the *Critique of Pure Reason* has shown, Kant only turns to discuss methodological concerns after reason has delineated and familiarized itself with its legitimate elements, faculties and forces. Following the *Transcendental Doctrine of Elements*, Kant's treatment of method is thus aimed directly against Descartes' *Discourse on Method* and the philosophical practices that take their cues from Descartes, who sought to identify the true method and criterion for truth at the outset. In the *Critique of Pure Reason*, method is instead to come last. The post-position of method harbours a radical critique and an attempt to break away from a long history of philosophizing, from Descartes to Leibniz until Kant's own teacher Christian Wolff, which as a whole is based on the difference between mathematical and philosophical modes of reasoning.

The *Discipline of Pure Reason* distinguishes between two ways in which theoretical reason puts its elements to use. According to Kant, these two ways are not equally intrinsic and proper to philosophy, however. On the one hand, there is the properly philosophical method, which legitimately belongs to the domain of philosophy. And on the other hand, there is the geometrical-mathematical method of construction, which essentially belongs to the domain of mathematics rather than philosophy. Differently to philosophy and its metaphysical objects of being – God, the soul and the world – mathematics concerns itself only with concepts of magnitude. As such, the central difference between philosophical and mathematical concepts lies in the fact that when it comes to mathematics, its concept 'already contains a pure intuition in itself'.[26] Consequently, mathematical reason can legitimately move from a concept to its corresponding object in pure intuition via construction, just like a geometer who ventures to draw a triangle. To a priori construct a concept in intuition means to supply it with its corresponding intuition through a priori presentation (*Darstellung*). Such an a priori presentation would be entirely illegitimate if the concept had any relation whatsoever to a possible content of intuition, since such a content would always have to be given empirically, and thus a posteriori. Since mathematics, however, concerns itself only with the pure a priori forms of intuition, which are already contained in a concept of magnitude, it 'gives the most resplendent example of pure reason happily expanding itself without assistance from experience'.[27]

Hoping to share in the apodictic certainty of mathematics' pure a priori mode of expansion, Kant describes how philosophy, in its own drive towards expansion, tries to adopt the mathematical-geometrical method for its own expansions. However, Kant vehemently argues against the adoption of this externally derived method by philosophy, as he claims that 'the mathematician can build nothing in philosophy except houses of cards, while by means of his method the philosopher can produce nothing in mathematics but idle chatter'.[28] In attempting to construct according to the geometric-mathematical method, meaning by means of concepts alone, philosophy aims to build a 'tower that would reach the heavens'.[29] But in fact, in (mis)using logic as an organon, philosophy will be left with nothing but a 'house of cards'. This is the case, Kant argues, since, in distinction to mathematics, there are no such things as definitions, axioms and demonstrations when it comes to philosophizing. First, only mathematical reasoning is said to deal with apodictically certain definitions, meaning that 'the concept is first given through the definition, it contains just that which the definition would think through it'.[30] Philosophy, on the other hand, cannot properly speaking define its concepts a priori. Instead, it must restrict itself to analytically dissecting concepts into their elements. And, differently than deriving from definitions as in mathematics, philosophical concepts only gain their validity via their adequacy to objects, which in any case always have to be given empirically.

Second, and concerning axioms as intuitive principles that are immediately certain a priori, philosophy has, strictly speaking, no single principle 'that deserves the name of an axiom'.[31] This might come as a surprise since the first principle in the *Analytic of Principles* carries the name *Axioms of Intuition*. However, Kant argues that 'the principle that was introduced there was not itself an axiom, but only served to provide the principle of the possibility of axioms in general, and was itself only a principle from concepts'.[32] Differently to mathematics, philosophy does not have intuitive principles (*axioms*) but only discursive principles (*acroams*), which must always be deduced and can never be a priori commanded.

The third and last point concerns demonstrations as apodictic, intuitive proofs. When it comes to philosophy, Kant writes that 'experience may well teach us what is, but not what could not be otherwise. Hence, empirical grounds of proof cannot yield apodictic proof'.[33] Consequently, only mathematics is said to contain demonstrations proper since its cognitions are derived from the construction of concepts in a priori intuition. Philosophy, on the other hand, is dependent on 'acroamatic (discursive) proofs' when seeking to secure the objective validity of its principles. As such, its principles are subject to discussion and discursive challenge, 'conducted by mere words (the object in thought)'.[34]

What becomes clear, then, is that Kant's critique of philosophy's employment of the geometrical-mathematical method is one and the same as his critique of reason's drive to expansion by means of the instrument of logic alone. To employ the geometrical-mathematical method in philosophy essentially means to (mis)use logic as an instrument of expansion. Relating Kant's critique of reason's use of logic as an instrument with which the *Critique of Pure Reason* opens and from which Kant's critical project as a whole assembles itself, to his critique of the history of philosophical method with which the *Critique of Pure Reason* closes, has two important consequences. First, it shows that Kant reads the history of metaphysical methodology as the history of instrumental reason. Second, however, it also shows that the *Critique of Pure Reason* as a whole and from its outset already performs a critique of philosophical methodology and thus a critique of the technical part of logic and its instrumental history. This is the case despite the fact that the discussion of the technical part of logic is relatively brief in comparison to the longer and more prominent *Doctrine of Elements*, as well as following after it. What remains to be shown, then, is whether the *Transcendental Doctrine of Method* will succeed in remedying reason's instrumental use of logic, or whether it will, in a similar way to the *Transcendental Doctrine of Elements*, take recourse to yet another technics.

Instrumentality, technics and the problem of mania

Kant's outright rejection of the geometrical-mathematical method has most certainly been underappreciated in Heidegger's inclusion of Kant in the movement of modern metaphysics towards its fulfilment in modern technics. As the previous chapter has shown,[35] Heidegger argues that this movement can be best observed in Kant's *Transcendental Analytic*, which laid open the a priori principles of the thingness of things, of how things can come into view in the modern period, articulating the very moment in which metaphysics 'became' modern technics.

The preceding discussion on Kant's *Critique of Pure Reason* has traced that and how Kant's own twofold critical intervention had itself been aimed against reason's instrumentality. The first book of the *Doctrine of Elements* is constituted against instrumental reason as it aims to lay down reason's proper bounds (famously claimed to be experience) and restricts logic from organon to canon. The second *Doctrine of Method* follows on by asking what reason can legitimately do with the elements delineated in the first book, aiming to legislate a properly philosophical method over and against the employment of the geometric-mathematical method in reason's instrumental drive towards expansion and becoming 'master over nature'.[36] The Kantian critical intervention thus ultimately and explicitly aimed to restrict reason in its instrumental intention of becoming metaphysical 'master over nature'.

We are thus faced with a complicated situation. On the one hand, the preceding discussion aligns Kant and Heidegger in their instrumental critique of philosophy, meaning that Heidegger radically underappreciated the reflections contained in the *Doctrine of Method*. On the other hand, however, and as I will show in the following, Kant's critical restriction of the technical *Doctrine of Method* introduces yet another notion of technics, a positive technics, due to the fact that reason's instrumental predicament is far worse than initially assumed. The current state of reason, Kant writes, is not only characterized by its problematic drive for expansion, whereby it pushes itself to instrumentalize logic in its attempt of constructing a tower of reason by way of the problematic geometric-mathematical method. Theoretical reason is furthermore afflicted with a pathological, persistent *mania* for speculative expansions (*spekulative Erweiterungssucht*).[37] The characterization of reason as *süchtig* engages the relation between reason and its expansions in a pathological setting in a twofold manner.

First, reason's pathological, persistent desire for expansion is due to the fact that while reason's metaphysical questions are said to arise from the very nature of universal reason itself, the consequent act of expanding cannot be said to come deliberately. To be subject to a persistent and pathological desire for speculative expansion means that reason does not freely choose to expand but that it compulsively expands whether it wants to do so or not, simply because it cannot but do so. In *Anthropology from a Pragmatic Point of View*, published in 1798 and compiled from lecture notes on anthropology delivered by Kant at the university of Königsberg between 1772 and 1796, Kant refers to various manias (*Süchte*) in his discussion of the four levels of the faculty of desire: propensity (*Hang, propensio*), instinct (*Instinkt*), inclination (*Neigung, inclination*) and finally passion (*Leidenschaft, passio animi*). Passions, Kant writes, are the most severe forms of 'inclination that can be conquered only with difficulty or not at all by the subject's reason'.[38] This is the case because they give rise to a number of manias (*Süchte*) that pathologically and persistently push us to purposively strive for objects such as honour, revenge, dominance, and the like, which remain, however, forever removed and thus out of one's reach. The difficulty of dealing with passions and their persistent, pathological expressions under the name mania (*Sucht*) lies in their ambiguous proximity to reason. Passions, Kant writes, 'can be paired with the calmest reflection, [. . .], they are not thoughtless, like affects, or stormy and transitory; rather they take root and can even co-exist with rationalizing'.[39] As lasting inclinations, it is thus essentially uncertain whether one can ever really rid oneself of passions. Passions are like poison that cause permanent illness, Kant writes. They do not admit of a cure but only of palliative care.[40]

Second, however, as lasting inclinations, the mind forms principles on them, involving a level of deliberation and rational action, due to which they take on a problematic semblance of reason. Passions take on 'the appearance of reason; that is, they aspire to the idea of a faculty connected with freedom, by which alone ends in general can be attained'.[41] Passions and their various manias thus engage reason in deliberations about the means best suited to reach an end, with the end itself, however, remaining 'prescribed [. . . by] inclination'.[42] As such, it would, however, be more accurate to characterize the deliberation involved in a passion as a mere quasi-deliberation and thus a means of plotting about how to best achieve an end set by inclination.

Third, and perhaps most importantly, passions take on the appearance of reason in their aim to acquire all possible means that might lead to its end. As such, they extend beyond particularity towards a quasi-universal reach. There remains, nevertheless, a crucial difference between the quasi-rational appearance of passions and universal reason. Passions 'aim merely at the possession of the means for satisfying all inclinations which are concerned directly with the end. [. . .] Possessing the means to *whatever* aims one chooses certainly extends much further than the inclination directed to one single inclination and its satisfaction'.[43] As such, despite the ambiguous proximity between the passions and rational-purposive action, Kant proposes that the two can be differentiated according to the strict criterion of universality. Concerning universal reason, both means and ends are subject to deliberation and rational choice.

Concerning the problem with reason's pathological and persistent mania for expansion it is of course reason itself, which is here said to be afflicted. What Kant seems to be proposing, then, is that reason is fixated and obsessed, so to speak, with its ends of general and special metaphysics. This means that if metaphysics follows its pathological and persistent mania for expansion, the resulting philosophical practice only appears like a rational undertaking, when in fact it is merely quasi-rational. These pathological ends are further said to 'arise from the nature of universal human reason'.[44] In the *Anthropology*, Kant distinguishes between innate and acquired passions. While innate passions are specified by Kant as the passions for freedom and sex,[45] the second kind of passions, acquired by culture, refer to the before mentioned manias (*Süchte*) for honour, dominance and possession.[46] When it comes to the *Critique of Pure*

Reason and reason's mania for speculative expansion, the metaphysical ends of reason (the questions of general and special metaphysics) are said to arise from within universal reason itself by Kant. Reason, Kant writes, is 'burdened with questions which it cannot dismiss, since they are given to it as problems by the nature of reason itself, but which it also cannot answer, since they transcend every [faculty] of human reason'.[47] In the end, reason has no one but itself to blame for its vexing dilemma. The source of the problem is located in reason's nature itself.

The B-Edition of the *Critique of Pure Reason* describes the problematic in the following way: metaphysical inquiry into the ideas of God, the immortality of the soul and freedom, is indeed a 'natural predisposition [*Naturanlage*]',[48] meaning that human reason naturally engages, always has and always will engage in metaphysical inquiry. It is naturally predisposed to and inclined towards metaphysics. However, if reason is to overcome what has so far been mere 'groundless assertions',[49] metaphysics must be raised from a natural predisposition, and thus from the level of its pathological, persistent mania, to the level of a free science. It is in this sense that the explicit aim of the *Critique of Pure Reason*, circling around the question 'How is metaphysics possible as a science?'[50] must be understood. This very process of transition is to be brought about by the *Critique of Pure Reason*. How, then, is this transition to be achieved?

The disciplining of reason

The problem that reason's pathological and persistent mania poses to Kant's critical project is that the boundaries between reason's 'proper' elements and its 'other' are engaged in a much more difficult composition than initially imagined. The *Anthropology* articulates the difficulty of distinguishing between inner and outer when it comes to passions and manias by way of the metaphor of the chain (*Kette*) on two occasions. First, Kant writes, 'no human being wishes to have passion. For who wants to have himself put in chains when he can be free?'[51] Second, Kant expands on the same metaphor a few pages further when he writes, 'because reason still does not ease off with its summons to inner freedom, the unhappy man [of passion] groans in his chains, which he nevertheless cannot break away from because they have already grown together with his limbs, so to speak'.[52] Returning to the problem of reason's mania for instrumental expansions with the metaphor of the chain in mind points to just how difficult it is to distinguish between reason's proper and improper elements.

While the instrument of logic initially allowed reason to expand its faculties and competences and thereby its bounds, reason is at the same time said to have become dependent and in a sense 'addicted' to its chains, it has grown together with its 'other' to the point of making the distinction between reason proper and supplement/instrument impossible. Consequently, it is one and the same instrument that allows reason to expand itself while also binding and enchaining it. Reason is bound to its instruments: the *organon* as instrument has become an organ in the modern biological sense.

The problem that Kant has to confront as a consequence of reason's 'addiction' is subsequently twofold. On the one hand, the problem remains to be one of a careful delineation of reason's legitimate elements and their proper bounds, which is the responsibility of the canon. On the other hand, however, reason's 'addiction' to the instrument of logic reveals that the careful delineation of reason's proper elements undertaken by the *Doctrine of Elements* would never have sufficed to change reason's mode of conduct. Simply delineating and 'knowing' about reason's legitimate elements and its proper bounds are not enough in the face of the severity of the problem. Knowledge alone seems to be powerless here since reason cannot but impulsively (ab)use logic as an instrument, precisely because it is addicted. Since most of the *Critique of Pure Reason* is filled with its considerations of elements, faculties and forces, and thus problems proper to the canon, it has hitherto appeared that method might be somewhat of an afterthought and indeed of secondary importance to the Kantian critical endeavour. For most of the *Critique of Pure Reason* it appears that the canon suffices in order to change reason's instrumental and problematic, constructive mode of conduct and bring metaphysics 'on the secure course of a science'.[53] Due to reason's mania for speculative expansion, the *Transcendental Doctrine of Method* now introduces an altogether new discourse according to which reason is to be subjected to a discipline in order to overcome its problematic method of construction by means of instruments. This discourse can be found in the section entitled *Discipline of Pure Reason*, which contains Kant's argument that, in fact, the discipline is to work on reason's inner nature, subjecting reason and its natural dispositions to a programme of rehabituation, which will in a second step allow reason to proceed according to the rules laid out by the canon. The discipline is to work on reason's ability or skill in following the rules laid out by the canon, whereby reason is to rise from its natural state of inclination, mania and ultimately addiction to the state of the free science of metaphysics.

In working on reason's ability, the second book of the *Critique of Pure Reason* is now the theatre in which the specifically German philosophical meaning of *Technologie* as *Kunstwörterlehre*, denoting problems of terminology and systematization, meets the notion of general art (*Kunst*) that derives from *Können* (ability, skill), which was previously introduced in the discussion of the art of schematization. Rather than employing the term *Kunstwörterlehre* like Walch, where it denotes the terminology specific to a domain, be it the sciences or philosophy, Kant speaks of method here as a *logische Kunst* (logical art). Employing the term *Kunst* as well as *Technik*, rather than *Kunstwörter* (art words, technical terms) or *Kunstwörterlehre* (the doctrine of technical terms), consequently indicates that for Kant method does not only denote a meta-discourse on *Kunst* and *Technik* but refers to the general notion of art (*Kunst*) itself, designating the practical ability [*Fertigkeit*][54] to proceed and bring forth according to rules.

The discipline in question is meant to work on reason's 'talent'[55] and skill of proceeding according to rules. Kant thus asserts that reason is subject to discipline and formation [*Bildung*], just like any other talent 'that already has by itself a tendency to expression [*Antrieb zur Äußerung*]'.[56] Encountering the word root *trieb* (drive) once again, which the translators of the Cambridge Edition have here chosen to translate as tendency, *Äußerung* (expression) refers to reason's technics of bringing forth logical systems (objectivation).[57] What is to be disciplined and subjected to formation is thus reason's tendency to expression (*Äußerung*) and its compulsion to step beyond and outside itself in its attempt to construct a house of reason. The resulting system of reason at stake here must as a whole be understood as an objectivation of the transcendental *Technik* of theoretical reason.[58] As mentioned earlier on, Kant employs the Aristotelian distinction between science (*episteme*) and technics or art (*techne*) in the *Critique of the Power of Judgment* in the sense that there is a fundamental difference between what one can do and what one knows. 'Only that which one does not immediately have the skill to do even if one knows it completely belongs to that extent to art.'[59] When Kant thus speaks of method as a logical art – the properly technical part of logic – he refers to that aspect of logic which 'one does not immediately have the skill to do even if one knows it completely'.[60] Conjoining this notion of *Kunst* with the adjective *logisch* (logical) in the context of the *Critique of Pure Reason*, the notion of art at play here is once again not determined in distinction to the systematicity of science and neither restricted to the mere domain of material production. As an explicitly logical art, it also designates theoretical reason's ability (or inability) to follow rules and put forth logical systems according to these rules. It is thus theoretical reason's *Technik*, its talent and skill of proceeding according to rules, which is subjected to a disciplinary formation towards the end of the *Critique of Pure Reason*. The disciplinary formation in question is differentiated by Kant from culture because culture produces skills additively 'without first cancelling out another one that is already present'.[61] What this means is that while culture works towards the formation of ever more skills in addition to one another, discipline works on an existing skill itself with the aim of changing it in the sense of both diverting its ends and shaping the very skill of achieving these ends. Discipline, Kant writes, wields 'the compulsion [*Zwang*] through which the constant propensity to stray from certain rules is limited [*eingeschränkt*] and finally eradicated [*vertilgt*]'.[62] As such, there are two consecutive steps when it comes to discipline. At first, discipline only ever imperatively says 'No!' to reason's natural disposition and its drive of expansion and objectivation (*Antrieb zur Äußerung*).[63] The first step of discipline is thus to whip reason into line. The martinet (*Zuchtmeister*)[64] doing the whipping is reason itself, which means that reason subjects itself to the practice of self-disciplining in order to shape its technical skill of proceeding according to rules. And while it is certainly 'humiliating for human reason that it [. . .] requires a discipline to check its extravagances [. . .]', the fact 'that reason can and must exercise this discipline itself, without allowing anything else to censor it, elevates it and gives it confidence in itself'.[65]

In a second step then, reason is said to undergo a change. As a result of being disciplined, reason no longer seems to have the same compulsive drive to step outside itself. Its previous pathological and persistent mania for expansion and objectivation via the instrumental use of logic has been cancelled [*aufgehoben*]. It has, Kant writes, been exterminated [*vertilgt*].[66] Differently to the *Anthropology*, where passions could only ever be dampened and admitted to palliative care, the *Critique of Pure Reason* proposes a more optimistic outlook on reason's passions. Kant seems to suggest that reason can indeed be successfully disciplined, through which the transition from metaphysics as a mere mania to a science proper can be accomplished. After being disciplined, reason is said to want what it should have, namely to proceed according to the laws, limits, and boundaries determinately laid out by the canon. Reason, famously, 'is not like an indeterminably extended plane, the limits of which one can cognize only in general, but must rather be compared with a sphere [. . .] from which its content and its boundary can also be ascertained with certainty'.[67] Kant's point here is thus that after reason's technical talent has been disciplined, the sphere (of experience) will have become reason's proper bounds and thereby its adequate object. Reason will no longer aim to expand itself beyond its proper bounds by means of the instrument of logic. The German word *abrichten* captures the double meaning of disciplining that Kant is referring to. At once denoting the compulsive measures of disciplining and habituation, it also evokes the work of re-orientating one's view and the change of perspective that necessarily comes with it. To be disciplined means precisely to rearrange and thus change the orientation of one's ends.

After reason's technical talent for proceeding according to rules has been disciplined and the art [*das Können*] of following the rules set by the canon has been acquired, reason's horizon will be congruent with the sphere of experience. What reason will want will be one with what it can and should have. Its transcendent mania to step outside itself via the instrument of logic will have been eradicated, and reason will be fully content with immanence; it will want only what it can have in terms of what it is legitimately capable of. Philosophy as a whole, Kant tells us, 'consists precisely in knowing its bounds',[68] at once pointing to Kant's general Epicurean outlook on philosophy while also referencing Rousseau, for whom 'a truly free man wills only what he can do, and does what pleases him [*ne veut que ce qu'il peut, et fait ce qu'il lui plaît*]'.[69]

7

The many faces of technics in the critical framework

The point of departure of this chapter, which began my investigation of the role and meaning of technics for Kant, was the identification and discussion of the general problem that motivated Kant's *Critique of Pure Reason* in the first place. I have argued that this general problem was reason's illegitimate and instrumental mania for expansion. Kant's starting point was thus nothing less than reason's instrumentality, with Kant arguing that reason compulsively employed logic as an instrument to expand itself towards places far removed from empirical experience so as to gain access to metaphysical questions or to its supposed real-world referents. Instrumentality and the corresponding geometric-mathematical method of construction were shown to lie at the heart of reason's diagnosed predicament, prompting the Kantian critical intervention as a whole.

The consequent discussion of the critical framework, which aimed to place this unruly, instrumental, and ultimately pathological reason into a restricted, critical economy, was organized according to the twofold route proposed by the two books of the *Critique of Pure Reason* as a whole. In the process of tracing both of these routes, it became clear, however, that each of them itself took recourse, in one way or another, to a technical remedy.

The canon was discussed as the first half of the critical remedy to reason's mania for expansion by means of the instrument of logic, and the primary question posed to it was whether, with regards to reason's elements, it indeed succeeded in averting the intrusion of the instrumentality of technics. Heidegger, Stiegler and Schlosser were all shown to have argued that, while the canon was intended as reason's path to emancipation from logical instrumentality, Kant nonetheless articulated nothing else but an implicit technics. According to Heidegger, the canon harboured Kant's silent articulation of enframing (*Gestell*), and thus the essence of modern technics. According to Stiegler, the canon silently relied on a number of technical substrata in order to secure reason's non-instrumental course of conduct, which Kant believes to have analytically separated reason from. The subsequent elaboration on Schlosser considered both Schlosser's critique, charging Kant with reducing cognition to a mere form-giving manufacture, as well as Kant's answer to the charge. While the term *Technik* was not explicitly employed in this debate by either Schlosser or Kant, I argued that their encounter nevertheless circled around what was already a technical problematic. The problem that I articulated in the following was that, ultimately, Kant characterizes schematization as an art (*Können*). Consequently, it is excessive to knowledge. It proceeds by way of the technical skill of forming by hand (*Handgriff*) and is thus quite literally a manufacture. In the last instance, Kant's canon of transcendental logic thus derives its self-proclaimed superiority over general logic by means of the art of schematization, and thus from a set of technical-manual *Handgriffe*, which, in the last step, are meant to secure the ground for judgment and the knowledge claims that flow from it.

The second path of critically restricting reason's instrumental (ab)use of logic looked at the second book of the *Critique of Pure Reason*, the *Doctrine of Method*. It is in the context of method that Kant's first explicit engagement with technics as technics (*die Technik*) rather than a general notion of art (*Können*) can be found. My argument concerning the role and meaning of method in Kant's critical aim at restricting the (ab)use of logic as an instrument focused on the role of discipline. The problem that discipline responded to was that simply knowing and stating the laws (the canon) according to which reason is to proceed lawfully was said to not be enough to keep reason from following its drive to expansion by means of instruments. This was argued to be the case because reason was diagnosed by Kant as being 'addicted' to its instruments of expansion. Consequently, reason was to be subjected to a discipline in order to 'internalize', as it were, the law of the canon as its second nature. After this process of formation and habituation, reason's instrumental drive to stray from the law is supposed to be eradicated. What is to be formed and disciplined is reason's skill itself, that is, the talent of proceeding and bringing forth according to the canon: reason's technics.

In conclusion, there are two different strands of technical thought at play in the problematic of discipline and the formation of reason's very own 'talent' for proceeding according to rules. On the one hand, Seibicke has shown how Kant was both aware of and intervened in the German discourse of *Technologie* as the methodological problem of systematization and terminology, which derives from the medieval scholastic notion of *Technologia*. Kant was the first to have shortened the Germanized term *Technologie* into *Technik*, most likely in order to distance this notion of technics from Beckmann's newly founded science of *Technologie*. On the other hand, this notion of method as a logical art and technics also employs a notion of a general concept of art (*Kunst*) that derives from skill and ability (*Können*), as 'that which one does not immediately have the skill to do even if one knows it completely'.[1] Thus, when the *Doctrine of Method*, the technical part of logic, subjects reason to a discipline, what is to be rehabituated is reason's inner nature itself, its practical capacity, general skill and ability (*Können*) to follow the rules and respect the boundaries set by the canon.[2] Only in this double-coding of technics as both systematization and ability (*Können*), can the Kantian engagement with method, addressed in the second book of the *Critique of Pure Reason*, be properly understood.

However, with a total of only six explicit mentions in the *Critique of Pure Reason*,[3] the majority of which are to be found in the *Doctrine of Method*, my inquiry into the role and meaning of technics has been severely restricted so far. It is in Kant's practical and aesthetic philosophy that technics will finally emerge from its shadowy existence, as the general notion of art as ability and skill (*Kunst* and *Können*) is aligned with and explicitly called *Technik* from then on. At the same time, the open contradiction between the systematic technics internal to methodological considerations and the non-systematic technics of art as general skill continues to persist and is subject to further renegotiation. The following chapter will critically trace how the positive notion of technics that Kant relied on in order to secure the critical restriction of reason's instrumental use of logic is subjected to further, explicit development from the *Groundwork of the Metaphysics of Morals* until the *Critique of the Power of Judgment*.

Part III

Technics between theory and practice: The critique of the power of judgment

The preceding chapter, on the role and meaning of technics in Kant's *Critique of Pure Reason*, presented a twofold picture. First, as the *Doctrine of Method* showed, Kant was familiar with and engaged the medieval scholastic notion of *Technologie*. Shortened to *die Technik*, the concept was employed by Kant to refer to the methodological task of the systematization and organization of science. Method was explicitly called a 'logical art [*Kunst*]', providing the technical part of logic.[1] This logical art was then shown to be subject to a double determination as both systematization and ability (*Können*), as reason's talent or skill of systematization was subjected to a discipline. Second, the canon, providing the counterpart to method, also turned out to harbour a technical problematic itself. While explicitly designed as the remedy to the problematic, instrumental organon of logic, the canon was itself shown to rely on the art (*Kunst*) of schematization and, thus, quite literally the technical skill (*das Können*) of grasping and unifying by hand (*Handgriff*). What was at stake, then, was ultimately a technics of pure theoretical reason.

In his 1785 *Groundwork of the Metaphysics of Morals*, Kant discusses technical imperatives as a problem of practical philosophy. In the subsequent *Critique of Practical Reason* (1788), Kant will, however, explicitly bring these same imperatives back into the jurisdiction of theoretical philosophy, raising the question of whether the re-alignment of the technical-practical with theoretical reason does not simply correspond to the distinction between pure and applied philosophy. Such was Bernard Stiegler's reading, for whom Kant's notion of the technical-practical designated the mere application of theoretical knowledge.

I will show in the following, however, that the *Critique of the Power of Judgment* (1790) affirmed that technics is indeed the responsibility of theoretical rather than practical philosophy. In the second part of the *Critique of the Power of Judgment*, Kant elaborated on a transcendental rather than empirical technics, which 'could have always been appended to the theoretical part of philosophy'.[2] The resulting idea of a technics of nature is charged in the *Critique of the Power of Judgment* with the task of orienting and securing the understanding and its knowledge claims, whereby acknowledging that the aporia of judgment could not be overcome by schematization alone. What in the previous chapter was discussed as Kant's problematic recourse to technical skills and a hidden art of grasping by hand is now explicitly discussed by Kant under the name of a transcendental technical power in the *Critique of the Power of Judgment*. As the following discussion will show, finally, it is neither nature nor our judgments about it, which are properly speaking technical. What is 'properly technical',[3] Kant writes, is our power of judgment itself.

8

The system of philosophy and the place of technics within it

The *Critique of Pure Reason* sets off from a preceding distinction of reason into theoretical reason on the one hand and practical reason on the other. Within that preceding distinction, its primary task was to deal with pure theoretical reason. In 1785, with the *Groundwork of the Metaphysics of Morals*, Kant began his investigation of pure practical reason, the investigation of which he continued in the *Critique of Practical Reason* (1788). The *Groundwork of the Metaphysics of Morals* prefaces its engagement with practical reason with the following, systematic reflection on the ancient Greek classification of philosophy in general: 'Ancient Greek philosophy was divided into three sciences: physics, ethics, and logic.'[1] While general logic, for Kant, is to function as a canon for the understanding and reason as the pure form of thought in general, both physics and ethics each have their own proper objects as well as sets of laws, belonging to what Kant calls material rather than formal philosophy.[2] Physics sets out to determine the laws of nature, while ethics lays out the laws of the will as a causal actor within nature. Since each of these two sciences can have an empirical as well as a pure a priori part, Kant concludes that metaphysics, as the pure a priori investigation of a determined set of objects, is in fact a twofold affair. On the one hand, there is a metaphysics of nature and, on the other hand, a metaphysics of morals, the latter of which 'has to examine the idea and the principles of a possible pure will and not the actions and conditions of human volition generally, which for the most part are drawn from psychology'.[3] Following this fateful distinction, the subject of Kant's investigation of practical philosophy consequently centres around the possibility of a pure will.

It was already in the preceding *Critique of Pure Reason* that Kant had offered a preliminary anthropological definition of the will when he wrote that 'in the human being there is a faculty of determining oneself from oneself, independently of necessitation by sensible impulses'.[4] Essentially, then, while the human is certainly in nature, one of its faculties is simultaneously not subject to said natural causality. The *Groundwork of the Metaphysics of Morals* moved on to extend this anthropological determination of the will to that of living beings in general 'insofar as they are rational',[5] meaning that the will is a power found in the human being as well as in possible other rational beings.[6] The causal power of the will is then discussed as twofold. First, a representation can become the determining ground of the will, which then works on bringing forth – in the sense of causally producing – the object corresponding to this representation in actuality. This is the case in all kinds of purposive actions where empirical objects are brought forth, like in the handicrafts or arts for example. The will is here determined by the determinate representation of, for instance, a table, in order to purposively put forth the means (wood, paint, screws, etc.) as well as make a table out of these means according to the representation of the end: a table. What in general is called technical production can here be seen to refer to the conscious end-realization through the determination of the will. The will is, however, also the faculty by which a subject determines its causality independently of one's physical capacity or technical expertise to do so or not, since the will is the power of self-determination in rational beings.[7] Both employments of the will are purposive. But in the first case, the will is determined by a concept belonging to the realm of nature and works on bringing about a corresponding object in actuality that is possible according to the laws of nature. In the second case, the causality is equally purposive, but the will is said to be free in the sense that it spontaneously determines itself. As such, the obligation originates within itself and has no end outside of itself, like in the first instance. Consequently, then, the determining concept of the will does not belong to the realm of nature or sensible appearance at all. It is rather the idea of freedom that is determining here, the objective reality of which as 'only an idea of reason' remains essentially uncertain.[8] And while the first mode of purposiveness is thus properly speaking the concern of psychology, the second mode of employment of the will is the concern of metaphysics proper.

Were this self-determination of the will to take place without any sensuous or worldly interference, then the will of such a (godly) rational being would be 'the capacity to choose only that which reason independently of inclination cognizes as practically necessary, that is, as good'.[9] Such a being would have no need for imperatives – for what it *ought* to do – since its will would fully coincide with the moral law of practical reason. The *Groundwork of the Metaphysics of Morals* proposes, however, that when it comes to the human being, there is only a partial overlap between will and reason, since the human is determined by both its worldly and intelligible existence. For Kant, this dichotomy or double determination between nature and freedom is an essential characteristic of the human being, who most certainly exists in nature and is thus subject to the heteronomous laws of nature while also standing under a wholly other set of laws, namely that of freedom, which are founded in reason only.[10]

Consequently, the human, for Kant, is dependent on a number of imperative 'oughts', which are the representations of objective principles that are compelling and obligating to our ambiguous human will. These imperatives are of two kinds. On one side, there are hypothetical imperatives, which represent 'the practical necessity of a possible action as a means to achieving something else that one wills'.[11] On the other side of this group of hypothetical imperatives, Kant situates just one counterpart, which represents an act as objectively necessary in itself without relation to any end – the categorical imperative.[12] While the categorical imperative commands objectively and universally, the hypothetical imperatives contain a certain level of contingency, as they are always oriented towards an end outside themselves and thus ultimately aim to satisfy a need or inclination. These imperatives represent a practical rule in relation to the will 'that does not straightaway do an action just because it is good'.[13] These two kinds of imperatives are then classified in a threefold scheme into 'rules of skill [*Regeln der Geschicklichkeit*]', 'counsels of prudence [*Ratschläge der Klugheit*]' and 'commands (laws) of morality [*Gebote der Sittlichkeit*]'.[14] The first of these, Kant writes, can best be called 'technical (belonging to art)',[15] the second pragmatic and the third moral.

Only five years after the *Groundwork of the Metaphysics of Morals*, the unpublished first introduction to the *Critique of the Power of Judgment* (1790) contains the following reference to the threefold classification of imperatives undertaken in the *Groundwork*:

> This is the place to correct an error which I committed in the *Groundwork for the Metaphysics of Morals*. For after I had said that imperatives of skill command only conditionally, under the condition of merely possible, i.e., *problematic*, ends, I called such practical precepts problematic imperatives, an expression in which a contradiction certainly lurks. I should have called them *technical* imperatives, i.e., imperatives of art.[16]

In a curious manner, Kant seems to have forgotten that in the *Groundwork* he had already termed the first kind of imperatives 'technical'.[17] Furthermore, in the intervening *Critique of Practical Reason* (1788), Kant again referred to them as technical, in the following passage:

> Propositions that in mathematics or physics are called practical should properly be called technical. For in these teachings it is not at all a question of the determination of the will; they only point out the manifold of the possible action that is sufficient to produce a certain effect, and are thus as theoretical as any proposition that asserts the connection of a cause with an effect. Whoever approves the effect must also be willing to approve the cause.[18]

Apart from insisting twice that the proper terminology for conditional imperatives as well as precepts from mathematics, is to call them technical, the above quotations bear witness to Kant's explicit attempt to reorganize the tripartite schema of the will into a twofold structure, which, according to the *Groundwork*, was still said to be technical, pragmatic and moral. The subsequent *Critique of the Power of Judgment* carries this attempt further by proposing that technical imperatives do not belong to the field of practical philosophy since, properly speaking, they concern the practical application of a concept of nature:

> If the concept determining the causality is a concept of nature, then the principles are technically practical [*technisch-praktisch*], but if it is a concept of freedom, then these are morally practical; [. . .] the former will belong to theoretical philosophy (as a doctrine of nature), while the latter will entirely by itself constitute the second part, namely practical philosophy (as a doctrine of morals).[19]

This point is repeated again in the later *Metaphysics of Morals* (1797). In reference to the preceding *Critique of the Power of Judgment*, Kant asserts that everything that is possible according to the laws of nature 'depends for its precepts entirely upon the theory of nature'.[20] From the *Critique of the Power of Judgment* onwards, the hitherto tripartite schema of the practical imperatives is consequently replaced by the twofold distinction into technical-practical (*technisch-praktisch*) principles on the one hand and moral-practical (*moralisch-praktisch*) principles on the other.[21] And while moral-practical principles belong to practical philosophy properly speaking since they concern the determination of the will and the concept of freedom, technical-practical precepts are explicitly said to belong to theoretical philosophy.

In *Technics and Time, 3* Bernard Stiegler argues that Kant considered technics to be a problem of mere application in his subordination of technics to theoretical philosophy *tout court*. Technics, Stiegler writes about Kant, 'is here indeed applied science: it has no opacity by *rights*; it can remain in the darkness of reason only *in fact* – by the *fact* of science's incompleteness'.[22] Understood as the mere application of science, Stiegler thus reads the Kantian notion of technics as essentially dependent on the state of the development of science. Technics is a mere stepping stone on the way towards the development of apodictic scientific knowledge, with Stiegler likening the Kantian relation between technics and the incompletion of science to the ancient Greek Aristotelian determination of *techne* in relation to the incompletion of nature (*physis*).[23]

Contrary to Stiegler's arguments, however, the following chapter will argue that the *Critique of the Power of Judgment* relies on a concept of technics that cannot be properly understood as the mere application of theoretical philosophy. I will show that from the *Critique of the Power of Judgment* onwards, Kant explicitly charges the technical with a transcendental function. As such, and in contrast to Stiegler's proposition, technics does not only *not* follow from science, but it is finally explicated and employed by Kant as what ultimately makes science possible.

9

The *Critique of the Power of Judgment* and the problem of nature's unruly things

Despite Kant's adamant claim to have enumerated, in the *Critique of Pure Reason* as a whole and in the *Analytic of Principles* in specific, all of the universal and necessary principles of the objectivity of objects, the *Critique of the Power of Judgment* deals with the problem of 'unruly things'. There are, Kant claims, some things that successfully resist the supposedly universal and necessary conditions of objects of experience put forward in the canon of the *Critique of Pure Reason*. In the face of these unruly things, the canon becomes powerless. Posing a problem in the properly philosophical sense, these unruly things are neither something, which would subject them to the canon, nor nothing, according to which we could treat them as ideas of reason. Upon encountering one of these unruly things, we are puzzled. But how is it possible to encounter such unruly things in experience and thus as part of nature, while it is at the same time impossible to a priori account for their possibility by means of the canon?

While indeed problematic within the order of knowledge, the *Critique of the Power of Judgment* proposes that there is a specific way in which one can nonetheless 'make sense' of these unruly things. Specified as living and beautiful things, it might not be possible to cognize them, but by introducing a difference between determining and reflecting judgment Kant argues that it is nevertheless possible to reflectively judge (*beurteilen*) these twofold unruly things according to an altogether different principle. This principle, he goes on to claim, is purposiveness (*Zweckmäßigkeit*).

Properly speaking, Kant offers two accounts of purposiveness. According to the first account of purposiveness, an end 'is the object of a concept insofar as the latter is regarded as the cause of the former (the real ground of its possibility); and the causality of a concept with regard to its object is purposiveness (*forma finalis*)'.[1] This first, objective account of purposiveness is complemented by a second, subjective account, according to which an end is the 'concept of an object insofar as it at the same time contains the ground of the reality of this object [. . .], and the correspondence of a thing with that constitution of things that is possible only in accordance with ends is called purposiveness of its form'.[2] Distinct from the principles of the understanding when it comes to determining judgment expounded in the *Analytic of Principles* of the *Critique of Pure Reason*, the subjective principle of purposiveness is what, according to the *Critique of the Power of Judgment*, now allows the power of judgment to reflexively judge (*beurteilen*) things in the absence of universal rules and principles supplied by the understanding, as it instead reflexively judges a thing as purposive for us. What this means is that, in a first step, upon encountering an unruly thing, the power of imagination can take up the manifold of what is given in intuition. However, this very first step already proves problematic from the perspective of the *Critique of Pure Reason*: has Kant not explicitly claimed in the *Critique of Pure Reason* that the manifold of what is given must essentially be orderable and placeable by the a priori forms of intuition, which are however themselves always already pre-ordered by the categorical order of the understanding?

The so-called A-Deduction[3] of the *Critique of Pure Reason* specifies that when it comes to cognition, the understanding enters third and last in the threefold synthesis.[4] The synthesis of recognition, performed by the understanding, renders the previous two modes of synthesis intellectual, as it is the concepts of the understanding that ultimately provide synthetic unity to a manifold of intuition. Thus, in the last instance, the synthesis of recognition must be said to precede even the ground provided by the power of imagination, which turns out to already essentially synthesize under the rule of the understanding in as much as the unity of the a priori forms of intuition is the effect of the understanding as well. Thus, while the synthesis of recognition is presented as the last and final step in the threefold synthesis, it is at the same time the 'transcendental principle of all the manifold of our representations (thus also in intuition)'.[5] In its function as first principle, the synthesis of recognition precedes the first synopsis of apprehension. This means, however, that the manifold of intuition, even in its pure and a priori form, is always already pre-shaped by the understanding. Sensibility's skill for ordering and arranging the manifold of sensation according to its a priori forms of intuition is thus in fact preceded by the understanding's act of ordering and arranging the a priori forms of intuition, and it is the understanding's prefiguration of the forms of intuition that makes the forms of intuition capable of ordering and placing the matter of intuition to begin with. This means that while the understanding is indeed dependent on intuition for objects to be given to it, the *Critique of Pure Reason* ultimately holds, at the same time, that only that which is thinkable by the understanding and its concepts can be given in intuition in the first place.[6] How, then, can Kant propose in his 1790 *Critique of the Power of Judgment* that one can nevertheless apprehend unruly things of beauty and the living in a manifold of intuition?

The crucial point here is that Kant seems to propose that unruly things can both be given to us and that a mode of apprehension different to the threefold synthesis laid out in the *Critique of Pure Reason* is possible. The problem posed by these unruly things goes to the heart of the a priori synthesis between intuition and understanding, and, as I will argue in the following, will shine a first light on the meaning and role that Kant designates to the technical. How, then, are unruly things taken up and what does technics have to do with it?

Despite having no corresponding concept, the 'form of a given object in empirical intuition'[7] can bring about an agreement between the first synthesis of apprehension and the third of 'presentation of a concept of the understanding (though which concept be undetermined)'.[8] This means that, rather than corresponding to any particular concept, the apprehended form of the unruly object corresponds to the form of concepts as such, their conceptuality, so to speak, such that it is in agreement with the faculty of understanding as a whole rather than with any concept in specific. This agreement between apprehension and presentation according to the forms of concepts as such is significant since it allows for a judgment to take place despite there not being any corresponding concept in specific. But if the judgment is not warranted by a determinate concept, allowing the power of judgment to subsume a manifold of intuition under it, what is the ground of this judgment?

Kant tells us that the apprehended form of unruly things effects nothing less than a certain kind of lucky relation between our cognitive faculties of imagination and understanding. This relation gives rise to a pure a priori pleasure, by which the mind (*Gemüt*) can feel itself.[9] We thus learn that reflective judgment allows the mind to feel itself. But the self in question here is nothing like the self-sameness of the merely logical and numerically singular transcendental unity of the 'I' that Kant introduced in the context of determining judgment. The feeling of self that Kant introduces in the context of reflective judgment allows for the *Gemüt* to feel itself alive and thus as an embodied being.[10] Kant claims that it is this feeling of self and its accompanying pleasure that provide the ground for reflective judgment. This means, says Kant, that the apprehended form of the unruly thing

> is related entirely to the subject, indeed to its feeling of life, under the name of the feeling of pleasure or displeasure, which grounds an entirely special faculty for

> discriminating and judging that contributes nothing to cognition but only holds the given representation in the subject up to the entire faculty of representation, of which the [*Gemüt*] becomes conscious in the feeling of its state.[11]

Something remarkable has thus occurred here. Despite the argumentative force of the *Critique of Pure Reason*, we can now see that, first of all, the mind (*Gemüt*) has dealings with a certain set of unruly things, the objectivity of which is not a priori determinable by the *Analytic of Principles*. Furthermore, the apprehension of these unruly things allows for a different access to the self than that provided by transcendental self-consciousness. Second, the feeling of self is crucial for another reason, since it is the ground of an 'entirely special faculty for discriminating and judging'.[12] This means that, despite not being able to make any objective knowledge claims about such unruly things, the feeling of life becomes the ground for a judgment nonetheless. It is the feeling of life which allows us to judge the given and apprehended representation as purposive (*zweckmäßig*) for us, meaning as purposive for our subjective cognitive faculties. What this ultimately means is that, faced with unruly things and in the absence of concepts of the understanding giving the law, nature 'specifies'[13] itself.

In its logical coding, species denotes the mediating term between genus and individual and thus corresponds to the schema, which sits between universal concept and particular object. This means that, in the absence of transcendental laws of the understanding, nature 'schematizes' or 'specifies' itself – it provides its own schema. It forms itself. And when this form is taken up by the imagination, the imagination finds itself in accord with the understanding. As already mentioned, this accord does not take place on the level of any specific concept and intuition, but on the level of the faculties themselves. The unexpected agreement between the faculty of imagination and the understanding reveals a structural adequacy of empirical nature to our cognitive powers as a whole. This structural adequacy might be revealed to us in reference to individual unruly things, which are then judged as beautiful. But Kant's interest in these unruly things expands further than any specific living or beautiful thing. As a matter of fact, the reflective judgment of unruly things 'prepares' the understanding to apply the principle of purposiveness to nature as a whole. Expounded in detail in the second part of the *Critique of the Power of Judgment* entitled *Critique of the Teleological Power of Judgment*, Kant moves on to argue that 'the self-sufficient beauty of nature reveals to us a [technics] of nature [*Technik der Natur*]'.[14] Thus, while the preceding chapter showed that and how schematization in the *Critique of Pure Reason* already relied on the technical-practical skill of grasping by hand, the *Critique of the Power of Judgment* picks up where the *Critique of Pure Reason* left off and introduces the idea of a technics of nature in reference to the unruly things of beauty and nature which, in the absence of concepts of the understanding, are said to schematize themselves.[15]

10

A technics of nature

The term *Technik der Natur* (technics of nature) was first introduced by Kant in the *Critique of the Power of Judgment.* It is a peculiar and idiosyncratic term, as the German philosopher Ulrike Santozki points out in her *Die Bedeutung antiker Theorien für die Genese und Systematik von Kants Philosophie.* Derived from a peculiar cross-reading of both the ancient Greek τέχνη φύσεως and the Latin *ars naturae*, both Santozki and Seibicke arrive in their linguistic and etymological analyses at the conclusion that there are no known predecessors for the Kantian term in the German history of philosophy.[1] Grammatically speaking, the term 'technics of nature' is a double genitive, a form employed by Kant throughout his writing. What this means is that the term can be read in a double sense depending on how one interprets the relationship between the two terms conjoined by Kant. First, the term can be read as referring to nature as the object of technics and thus to nature as subject to a transformation by technics. Second, the term can, however, also be read as referring to nature's technics and thus as the suggestion that nature has its own kind of technics, its own way of organizing and regulating itself. In the following, I propose to proceed systematically, considering not only the composite term but also the two relata in their own right that are put into relation in the composite term.

Nature

The *Critique of Pure Reason* had famously claimed that when considered materially, nature designates nothing but the 'sum of appearances, hence not a thing in itself but merely a multitude of representations of the mind'.[2] When considered formally, Kant continues, it is the 'order and regularity in [the appearances] that we call nature'.[3] Having been organized in relation to the one numerically identical and transcendental unity of apperception, nature's unity and systematicity were thus claimed to be cognizable entirely a priori.[4] Differently to this account of nature as the sum total of all possible objects of experience, the systematization of actual objects of empirical knowledge appears to be somewhat of an afterthought, which is nevertheless necessary due to reason's own demand for unconditioned completeness.[5] And while on the one hand, the *Critique of Pure Reason* suggests that particular laws are flowing unhindered from ever higher and universal law, Kant points out on the other hand that 'empirical laws, as such, can by no means derive their origin from the pure understanding [. . .]. But all empirical laws are only particular determinations of the pure laws of the understanding, under which and in accordance with whose norm they are first possible, and the appearances assume a lawful form'.[6]

This means that the *Critique of Pure Reason* ultimately left behind a problematic relation between its universal and particular laws, the dissonance of which it did not manage to overcome. While the universal and necessary formal requirements for any possible object of nature are indeed supplied by the understanding and thus secured a priori, the particular object itself must, in any case, always be supplied by actual experience a posteriori. From the perspective of the lawful understanding, the particular is then always considered a source of contingency rather than universality and necessity. The understanding can only anticipate those aspects of the object that the particular shares with the universal, leaving its differences untouched. The object's particularity is precisely that which the understanding can under no circumstances anticipate; it will always come as a surprise. In the context of the *Critique of the Power of Judgment*, Kant picks up this problem when he writes that 'the a priori derivation of the particular laws from the universal, as far as what is contingent in the former is concerned, is impossible through the determination of the concept of the object'.[7] Thus, on the one hand, nature is a priori subject to the necessary and universal laws of the understanding, while, on the other hand, there is a 'contingency of the agreement of nature in its products in accordance with particular laws'[8] in determining judgments.

Without mentioning that the subsequent supplementary systematization of the contingent empirical objects of experience, necessary in order to secure the relation between universal and particular laws of nature, had previously been assigned to reason in the *Critique of Pure Reason*,[9] the *Critique of the Power of Judgment* poses the problem of the contingency between universal and particular as a problem of and for the power of judgment itself. In its reflective use, the power of judgment is now tasked with showing how, and securing that, nature in its particularity, and not only in its universality, is indeed adequate to and, thus, purposive for our cognitive faculties. It is to this end that Kant engages with nature's unruly things and finally introduces his idea of a technics of nature. What the articulation of reflective judgment in the *Critique of the Power of Judgment* entails is thus nothing less than the reduction of the previously claimed to be sufficient and complete competence of the understanding 'to bring the existence of appearances under rules a priori'.[10]

The *Critique of Pure Reason* claimed to contain all of the constitutive principles for objects of experience. As such, it presented itself as fully capable of a priori dictating the universal conditions according to which undetermined appearances could become objects for us. From the point of view of the *Critique of the Power of Judgment*, it now appears that the all-powerful constitutive principles of the *Transcendental Analytic* are no longer sufficient by themselves to a priori secure the subsumption of appearances under universal forms. This means that it is not only unruly things that the constitutive principles of the understanding cannot seem to grasp; even when it comes to what were previously believed to be 'ruly things', the understanding now seems to be dependent on something further in order to, in the last step, secure the adequacy of the particular object of nature to our cognitive faculties. The term 'nature' in the newly introduced composite term 'technics of nature' thus does not refer to the concept of nature in its universality as stipulated by the understanding. It rather concerns nature in its particularity and actuality, and the problem of its adequacy and purposiveness for our cognitive faculties both in determining and reflective judgments. Kant's aim, then, is to complement the unity of nature, a possibility provided by the constitutive principles of the understanding, with the regulative idea of a technics of nature that provides the systematicity of nature in its actuality.

Technics

Facing 'nature' in the double genitive stands the complementary term 'technics'. We have already seen that nature refers to actual nature's contingency from the perspective of the understanding and the need to secure its 'system-adequacy'[11] by way of the power of judgment. The term 'technics' should, subsequently, also have something to do with systematicity. In the history of philosophy, the relation between technics and the system has, however, been notoriously difficult to reconcile. Aristotle famously distinguished his notion of *techne* from the necessity, universality, and thus systematicity of science (*episteme*). According to Aristotle, *techne* is situated between *empeiria* (experience) on the one hand and *episteme* (science) on the other, arising when '[. . .] from the many cases of thinking in experience a single assumption is formed in connection with similar things'.[12] As such, *techne* is differentiated from the systematicity of science in the strict sense, since for Aristotle science deals with what 'is of necessity in the unqualified way'[13] – the eternal. Even though in practice science and *techne* are sometimes used interchangeably by Aristotle, as, for instance, in the beginning of the *Nicomachean Ethics* where Aristotle refers to a number of *technai* as sciences,[14] *techne* and science are nevertheless theoretically and logically strictly separated. While *techne* deals with the domain of the variable, science concerns the domain of the invariable and eternal. Aristotle writes: '[there] is a sense in which [techne] and chance operate in the same sphere, as Agathon says: "[*Techne*] has a love for chance, and chance for [*techne*]."'[15] In contrast to the eternal, systematic objects of science, which can neither come into existence nor ever cease to be, technical objects thus ultimately belong to the field of contingency and are connected to chance.

We have already seen in the previous chapter that Kant draws on two parallel genealogies of technics. On the one hand, he employs the Aristotelian notion of *techne*, which also features in the *Critique of the Power of Judgment*, when Kant argues that there is a fundamental difference between technics and art on the one hand, understood as generalized skill and ability (*Können*), and knowledge on the other. Art, Kant writes, is 'distinguished from a theoretical [skill], as technique [*Technik*] is distinguished from theory (as the art of surveying is distinguished from geometry)'.[16] At the same time, however, Kant also employs the medieval scholastic notion of *Technologie* shortened to *Technik*. This notion is located internally to theoretical philosophy, as it explicitly refers to the methodological problem of systematization and terminology. It concerns nothing less than the skill of systematizing. However, the *Architectonic of Pure Reason* still affirmed in quasi-Aristotelian fashion that there are ultimately two forms of unity with which the system of a science can be built. The architectonic form of unity, Kant writes, is provided by reason entirely a priori, since reason is itself essentially architectonic: 'Human reason is by nature architectonic, i.e., it considers all cognitions as belonging to a possible system.'[17] Technical unity, on the other hand, is outlined 'empirically in accordance with aims occurring contingently'.[18] The *Critique of Pure Reason* thus ultimately affirms that science proper can never 'arise technically, from the similarity of the manifold or the contingent use of cognition *in concreto* for all sorts of arbitrary external ends, but arises architectonically, for the sake of its affinity and its derivation from a single supreme and inner end, which first makes possible the whole.'[19]

When it comes to the *Critique of the Power of Judgment*, Kant appears to have distanced himself from this previously upheld metaphysical distinction between science and technics and the two corresponding architectonic and technical forms of unity that one finds in the *Critique of Pure Reason*. Ulrike Santozki points out that the term 'technics', employed by Kant in the *Critique of the Power of Judgment* as, for instance, in the neologism technics of nature is testimony to the disappearance of Kant's earlier efforts of distinguishing between the architectonic and the technical.[20] She points out how one finds instances throughout the *Critique of the Power of Judgment*, where instead of speaking of an 'architectonic understanding'[21] or an 'architectonic of an intelligent world-author',[22] Kant could equally have employed the terms technical and technics, as well as art.[23] Santotzki's argument is thus that from the *Critique of the Power of Judgment* onwards, Kant employs the terms architectonic and technics interchangeably, illustrated lucidly by the example that Kant employs the expressions 'intelligent world-author' and 'the highest artist'[24] in one and the same passage.[25] Essentially, then, from the *Critique of the Power of Judgment* onwards, technics does not designate the mere application of theoretical notions and rules. And neither does technics designate the contingent counterpart of the systematic science. Rather, Kant introduces the idea of a technics of nature to designate the systematic self-specification of nature for the purpose of securing the system-adequacy of nature in its actuality.[26]

A hypothetical restriction

There is, however, an important restriction to the idea of a technics of nature. Kant writes:

> we shall in the future also use the expression '[technics]' where objects of nature are sometimes merely *judged as if* their possibility were grounded in art, in which cases the judgments are neither theoretical nor practical (in the sense just adduced), since they do not **determine** anything about the constitution of the object nor the way in which to produce it; rather through them nature itself is judged, but merely in accordance with the analogy with an art, and indeed in subjective relation to our cognitive faculty, not in relation to the objects.[27]

Thus, the idea of a technics of nature can and should be employed in relation to nature, but it must under no circumstance be mistaken as a constitutive principle for the possibility of certain products of nature or nature as a whole.

Instead, Kant writes, it is only a means for 'elucidation (exposition)'.[28] By way of judging nature technically according to the reflexive principle of a systematic ordering of nature in its particularity, we can proceed with the following presupposition: we can reflexively judge nature as if its forms were the purposive product of the representations of a synthetic universal and thus of God. What this ultimately means is that Kant arrives at the notion of a technics of nature through relativizing our discursive human understanding in contradistinction to the understanding of the Christian God of creation.

Human, discursive understanding, as the *Critique of Pure Reason* demonstrated at length, can only go from the universal to the particular, which is the very reason that the contingency of the particular first arises to begin with.[29] Complementing this human, discursive understanding, Kant places the intuitive understanding of the Christian God of creation, which is essentially independent of the need for both concepts and images.[30] This is the case because, as intuitive, the Godly understanding is said to be able to go directly from the intuition of the given whole as a synthetic universal to its particular parts because God immediately creates what he intuits. Consequently, God's representation of the whole contains 'no contingency in the combination of the parts'.[31] Kant's proposition is thus that the ground of the idea of a technics of nature, which is introduced by Kant to secure the systematicity of nature in its actuality, is nothing other than the representation of such a synthetic universal 'in accordance with the rule of ends'.[32] However, in being derived from a relativization of our human and discursive understanding in contrast to the intuitive understanding of God, the resulting representation of the synthetic universal is valid only with a hypothetical restriction. This means that we can only judge nature *as if* God had brought forth certain natural products in 'analogy with an art',[33] and thus in accordance with ends the representations of which are synthetic universals and thus systematic wholes, which can then be extended to nature as a whole.

While certainly subject to this important hypothetical injunction, the idea of a technics of nature nevertheless gives us the assurance 'that even with regard to its empirical laws nature has observed a certain economy suitable to our power of judgment and a uniformity that we can grasp'.[34] And while only designating a regulative ground for our power of judgment, this regulative ground must essentially always be presupposed in any judgment whatsoever, since its role is nothing less than the supplementation of the unity of nature in its possibility (provided by the understanding) with that of a systematic technics of nature in its actuality (provided by the power of judgment). What this ultimately means is that without this idea of a technics of nature, the understanding 'could not find itself in [nature]',[35] even when it came to determinate judgments. As such, the regulative idea of a technics of nature performs an essential orienting and grounding function for any judgment whatsoever, in the sense that it fills us with the assurance that we are indeed in the right place and equipped with the right faculties for grasping nature in both its possibility and actuality.

What the preceding line of argumentation has shown is that, essentially, Kant introduces the idea of a technics of nature in order to secure nothing less than the possibility of science as a whole. Due to the hypothetical restriction, however, the possibility of science in its systematicity is premised on a condition that cannot be objectively secured in the end. The hypothetical restriction of technics of nature as a mere regulative idea means that, while we can never be sure that nature has indeed been made 'for us', we must nevertheless proceed as if it was nevertheless the case.[36] Importantly, then, Kant's idea of a technics of nature in the end reveals more about us than about nature itself, since it is no one but ourselves who put the idea of technics into nature. 'We put, it is said, final causes into things, and do not as it were draw them out of their perception.'[37] There must then, in a way, be two kinds of technics at work in Kant's argumentation in the *Critique of the Power of Judgment*, both a technics of nature which works itself out on the surface with a hypothetical restriction as well as its underlying condition, which is what allows us to judge nature technically in the first place. The *Critique of Pure Reason* has been shown in the preceding chapter to take recourse to a transcendental technics as it secured schematization by way of the hidden art of grasping by hand or the technical part of logic. Does Kant's claim in the *Critique of the Power of Judgment* that it is ultimately us who put the idea of technics into nature rely on an underlying, transcendental, technical condition of our power of judgment?

11

Three kinds of technics

The preceding discussion has argued that the idea of a technics of nature, which Kant introduces to overcome the problematic relation between universal and particular laws that would otherwise threaten the possibility of science as a whole, is subject to a hypothetical restriction. This means that we can only reflexively judge nature *as if* there were a technics of nature. It is thus ultimately ourselves, Kant writes, who put the idea of finality into things (in the case of living things) and, by extension, technics into nature as a whole. What, then, is the underlying human faculty or power that allows us to judge nature technically?

The first candidate to take into consideration is the aforementioned notion of technical-practical purposiveness. Differentiated from science as a skill and ability (*Können*) as 'to be able from to know',[1] this notion of technics is purposive. Purposiveness in its technical instantiation designates the causality of a concept with respect to its object, with the concept functioning as the determining ground for the will to bring forth and produce the object.[2] Technical-practical purposiveness thus designates all kinds of purposive actions where empirical objects are brought forth, as in, for example, the handicrafts, referring quite literally to the application, or rather realization, of a concept of nature by means of technical-practical reason.[3] But, as Gerhard Lehmann has pointed out, there is a fundamental difference between the idea of purposive technics and the idea of a technics of nature. A purposive technical-practical act designates a conscious end-realization by means of the will. The idea of a technics of nature, however, merely designates the presentation, rather than the realization, of ends in nature.[4] As such, Lehmann points out that the two technical instances appear too heterogeneous to justify their analogous use without a third, mediating instance. And, indeed, there is a third concept of technics at work in the *Critique of the Power of Judgment*, which stands in a double analogy with the objective purposiveness of technical-practical acts and the purposiveness that ultimately leads to the idea of a technics of nature. It is this third technics which functions as the underlying condition of both.

The *Critique of the Power of Judgment*, as stated above, consists of two books. The first book deals with aesthetic judgment, while the second book is termed *Critique of the Teleological Power of Judgment* and deals with teleological judgment. While the first book shows that self-sufficient beauty 'reveals to us a [technics] of nature [*Technik der Natur*]',[5] it is only the second book, which deals with natural, living things and directly engages the idea of a technics of nature. In the second book, Kant proposes that upon encountering a natural, living thing, the power of judgment freely apprehends the form of the given object, just like in aesthetic judgment. However, differently to the judgment of beautiful things, which effects an agreement of the faculties rather than any specific concept and intuition, with regards to natural things, the power of judgment is said to compare the thing 'with reason and its principle of the possibility of a system, [. . .] if this form is found in the object, the purposiveness is judged objectively and the thing is called a natural end'.[6] In employing a systematic idea of reason as a measure (a representation of a synthetic universal), the power of judgment does not lawfully subsume particular cases under determinate rules of the understanding as it does in determining judgments, where the power of judgment is said to work just 'like an instrument'[7] according to the schematized rules of the understanding. Rather, teleological judgment is defined in explicit contradistinction to problems of mechanism, instrumentality and application. In comparing a given empirical thing with 'reason and its principle of the possibility of a system',[8] the power of judgment is instead defined as a properly technical power. What is technical, then, according to the first introduction to the *Critique of the Power of Judgment*, is neither nature, nor our judgments of it, but rather 'the power of judgment, on whose laws they are grounded, and in accordance with it we will also call nature technical'.[9]

What allows us to judge nature technically is thus the power of judgment. And this power of judgment is termed technical precisely because it measures and compares given things to reason and its systematic form rather than working under the law of the understanding. In the last step, this technical power of judgment thus leads us back to the faculty of reason, the discussion of which formed the first part of this chapter. How so? In a problematic sense, the technical power of judgment is free from a mechanical relation to the understanding like in determining judgment. On the other hand, it is not free in quite the same way as aesthetic judgments are, where 'no determinate concept of the object at all is required'.[10] And yet, no determinate judgment is generated either. Distinct from both determining and aesthetic judgments, the technical power of judgment appears to stand in a problematic relation to reason when it enacts teleological judgments, since it is both free from prior rule and gives itself a rule, while at the same time nevertheless referring to a measure derived from reason. In a strange way, then, the technical power of judgment gives itself a rule by comparing natural forms to reason and its systematic form.[11]

12

The return of the problem of technics

As the preceding pages have shown, the *Critique of the Power of Judgment* both places technics 'after' science as a problem of application while, at the same time, technics is also placed 'before' science, in the sense that the idea of a technics of nature, put forth by the technical power of judgment, is introduced in order to secure the possibility of science in the last step. Thus, while the *Groundwork of the Metaphysics of Morals* and the *Critique of Practical Reason* systematically aligned technical-practical reason with empirical philosophy of nature, due to which it did not need a critique, I have argued that the *Critique of the Power of Judgment* is the place in which Kant introduces a technical power, the technical power of judgment, as the common denominator for and as what ultimately makes both technical-practical action and the idea of a technics of nature possible. It is here that Kant's idiosyncratic combination of both the Aristotelian and Scholastic notions of technics most clearly stands forth. As proposed by Seibicke, Kant's notion of technics ultimately stands in close proximity to art (*Kunst*), deriving from capacity/ability (*Können*). The technics of systematization in logic, as well as all other systematizing and methodological forms including technics of nature, are themselves nothing but particular, objectivized forms of this practical capacity.[1]

Paradoxically, however, differently to the first part of the *Critique of the Power of Judgment*, which concerns the aesthetic power of judgment, the second book of the *Critique of the Power of Judgment*, which concerns itself with 'teleological' power of judgment, 'could have always been appended to the theoretical part of philosophy'.[2] Ultimately, then, the second book doesn't, properly speaking, belong to the *Critique of the Power of Judgment*. The investigation of this technical power, finally, could or perhaps rather should have been added to the *Critique of Pure Reason*, with the exact place of insertion remaining unspecified by Kant.

If this situation was not complicated enough already, it is troubled further by the fact that it is only the longer, first introduction to the *Critique of the Power of*

Judgment, most likely written before the second part of the *Critique of the Power of Judgment* on teleological judgment,[3] which names the power of judgment 'technical' in its comparison of particular natural things with the form of a system. The shorter and eventually published second introduction, as well as the main body of the *Critique of the Power of Judgment* itself, turns away from what the first introduction had termed technical power of judgment, referring instead to a teleological power of judgment. Only the term 'technics of nature' is retained in the final edition of the *Critique of the Power of Judgment*. Rather than locating technics in relation to the power of judgment, the second book of the *Critique of the Power of Judgment* does, however, still speak of reason in its technical use, when Kant writes that ultimately, technics of nature designates a 'kind of causality in nature, in accordance with an analogy with our own causality in the technical use of reason [im technischen Gebrauch der Vernunft]'.[4] Nevertheless, Kant will refer to the technical power of judgment again in the later published *Anthropology from a Pragmatic Point of View*, when he writes about the power of judgment as a 'faculty, which is aimed only at that which is feasible, what is fitting, and what is proper (technical, aesthetic, and practical power of judgment)[. . .]'.[5] Compiled from lecture notes to Kant's annual anthropology lectures held between 1772 and 1796 at the University of Königsberg before the final publication of the *Anthropology* in 1798, this reference indicates that Kant's intellectual engagement with a technical power of judgment was not stopped short with the terminological changes undertaken between writing the first and the second book of the *Critique of the Power of Judgment*, but rather continued after the completion of the *Critique of the Power of Judgment*.

Rather than reading the changes in Kant's terminology as a mere problem of presentation, I would like to propose a different reading. My argument is that both Kant's practical and aesthetic philosophies are testament to the development and dynamic internal to Kant's thinking of technics, which takes place across both his theoretical, practical and aesthetic philosophy. What I mean by this is the following. In the *Critique of the Power of Judgment*, nature in its actuality is called technical before admitting that it is, properly speaking, not. In the final print version of the *Critique of the Power of Judgment*, what was hitherto named the technical power of judgment as the technical instance proper is no longer called technical. What is now called 'teleological power of judgment' is furthermore said to not belong to the *Critique of the Power of Judgment* properly speaking, but should have been added to the *Critique of Pure Reason*. And what has already been called technical, the technical imperative, is insisted upon again and again by Kant that it should properly be called technical.

Kant's writings, from the *Groundwork of the Metaphysics of Morals* up to the *Critique of the Power of Judgment*, thus provide the stage for Kant's dynamic engagement with technics, at once reduced and pushed back to a problem of mere application of science while at the same time pulled out into the open in order to secure the possibility of science as a whole, at once explicitly called into question while at the same time silenced into obscurity. Technics has become a problem for the 'problem thinker' Kant.[6]

While both the systematic place and role of the technical power of judgment and the idea of a technics of nature ultimately remains instable from the perspective of the *Critique of the Power of Judgment*, the last, unfinished manuscript that Kant worked on from 1796 until shortly before his death in 1803, entitled *Opus Postumum*, takes as its starting point the transition (*Übergang*) between the metaphysical concept of matter and the physical concept of matter, and thus in effect continues the earlier work, addressed in the *Critique of the Power of Judgment*, of bridging nature in its universality and particularity. As Kant works on this transition, he encounters a cascade of problems, which finally lead him back to the critical philosophy that he initially believed to have concluded with the *Critique of the Power of Judgment* in 1790, when he proclaimed that 'with this I bring my entire critical enterprise to an end'.[7] As he mercilessly turns, at the very end of his life, to rethink the critical project anew, most if not all of the critical terms will be subjected to scrutiny, while technics, and specifically technical-practical reason, will be entrusted with an ever-increasing role within the system. As the following will show, what thus remains in unfinished form as the *Opus Postumum* is nothing less than the outline of what would have become Kant's philosophy of technics.

Part IV

Technics in the *Opus Postumum*

The preceding discussion has shown that and traced how technics occupied an important yet uneasy place within the *Critique of the Power of Judgment*. This next chapter will move on from the *Critique of the Power of Judgment* to the *Opus Postumum* (*Nachlasswerk*), the last, unfinished manuscript that Kant worked on from 1796 until shortly before his death in 1804. While Kant reportedly intended the *Opus Postumum* to provide the 'keystone' (*Schlußstein*) that would finally bring systematic closure to his doctrinal system,[1] the development of this keystone did not so much as effect the sought-after systematic closure as did it lead to the dynamic reworking of the system itself. Little would remain as it was, and, as I will show in the following, technics finally stepped out into the open and revealed, at last, its systematic role and position in Kant's system of transcendental philosophy.

I will begin this chapter by way of a brief introduction into the publication history of the *Opus Postumum* and the role played by Gerhard Lehmann in this context. Lehmann was not only one of the editors of the *Akademie-Ausgabe* of the *Opus Postumum* but also wrote, in his own philosophical work, about the concept of technics in Kant. The chapter will then continue with a discussion of technical-practical reason within the restricted context of the *Opus Postumum*. Technical-practical reason appears substantially for the first time in Fascicle VII, written between April and December 1800, and as a whole contains Kant's writings on self-positing, that is, a series of acts by which the subject makes itself into an object of experience. Intending to outline the context in which Kant's discussion of technical-practical reason both emerged and developed, I will undertake a reconstruction of Kant's account of self-positing in order to locate the role and responsibility assigned to technical-practical reason within it. In most instances, Kant argues that self-positing consists of two acts: a first merely logical and analytic act, which is then followed by a second, synthetic act, in which the subject posits itself as an object in space and time. I will argue that these two acts are themselves subject to a transition. And it is in and as this very transition between the first and second act of self-positing that Kant introduces reason in its twofold mode as technical-practical reason and moral-practical reason. The second act of self-positing, I argue, is determined by reason in its twofold form, due to which it takes on both a moral-practical and technical-practical form. In order to account for the intricacy of Kant's argument about technical-practical self-positing, the act by which the subject posits itself as an empirical object, it is necessary to outline Kant's renewed engagement with the doctrine of ideas, and in particular, the idea of world as the subject of cosmology. Following a discussion of the idea of world, knowledge of the world (*Weltkenntnis*), and philosophy in the world concept, I will show that technical-practical self-positing in the *Opus Postumum* always already takes place in reference to the idea of the world, first brought forth by technical-practical reason for the sake of the second act of self-positing. The idea of world, I will argue, is thus the first and most important product of technical-practical reason, forming at once the domain in which the human, as a technical-practical subject, appears to itself, as well as what in another register makes science possible. Cosmology becomes cosmo-technics in the *Opus Postumum*.

It is my contention that Kant's recourse to the technical as both technical-practical skill and the methodological systematization of logical and real systems, first encountered in the *Critique of Pure Reason* before being further developed under the name of technics of nature and the technical power of judgment in the *Critique of the Power of Judgment*, is now, in the *Opus Postumum*, subject to explicit and systematic development under the name of technical-practical reason. What in the *Critique of Pure Reason* appeared as the first tentative steps of Kant's engagement with a transcendental technical remedy to reason's (ab) use of logic as an instrument of expansion is now explicitly developed under the name of technical-practical reason.

13

Kant's *Opus Postumum* and Gerhard Lehmann

What is today known as the *Opus Postumum* (*Nachlasswerk)*[1] is a collection of 290 loose sheets and folios, consisting of 525 handwritten pages of manuscript that Immanuel Kant worked on primarily between the years 1796 and 1801, with some pages dating back as early as 1786 and some as late as 1803.[2] The manuscript contains what Kant reportedly hoped to be his '*Hauptwerk*' or '*Chef d'oeuvre*', and as such, it was anticipated by Kant to contain nothing less than the 'keystone to his entire doctrinal system'.[3] It is most often referred to by Kant as providing a transition from the metaphysical foundations of natural science to physics and consequnetly called 'Transition [*Ubergang*] from the metaphysical foundations of natural science to physics'[4] or 'Overstep [*Überschritt*] from the metaphysics of corporeal nature to physics'[5] in ever-changing variations. With the turning of each page of the manuscript, we accompany Kant in his daily practice of philosophizing until work on the manuscript ceased around February 1803 before his eventual death on 12 February 1804.

Following Kant's death, his executor Ehregott Andreas Christoph Wasianski handed over the unfinished manuscript for inspection to Johann Friedrich Schultz, who concluded that the manuscript was incomplete to the point of making an edition impossible, due to which it disappeared into relative obscurity for most of the nineteenth century. Due to rumours about Kant's state of mind towards the end of his life, it slowly became the object of much speculation.[6] In 1890, the *Preussische Akademie der Wissenschaften* commissioned Erich Adickes to edit and prepare Kant's *Handschriftlicher Nachlass* for publication in the encompassing academy edition of Kant's work. As part of his research, Adickes also inspected the *Opus Postumum.* During a four-week-long investigation of the manuscript in 1916, Adickes claimed to have established its chronological, genetic structure. Arguing that the very final Fascicle I was virtually complete by 1801, and thus at a time when Kant had still been in good health, Adickes advised the *Akademie der*

Wissenschaften to include the manuscript within its edition. However, due to a change of ownership of the manuscript, a situation arose in which the manuscript was going to be transcribed by a young assistant editor to Artur Buchenau at the publishing house De Gruyter, with Adickes functioning as a mere external editor. The name of the assistant editor was Gerhard Lehmann, a young student of philosophy who would later become known for his contribution to Kantian scholarship as both an editor and commentator of German philosophy with a specialization on Kant.[7] The relation between the editors of the *Opus Postumum* however proved to be difficult and soon deteriorated. Most importantly, Buchenau and Lehmann decided to steer away from Adickes's editorial principles that had hitherto been applied to all of Kant's *Handschriftlicher Nachlass*, which also meant that they did not adopt Adickes's chronology. Adickes himself critiqued the reliability of the transcriptions of the then twenty-three-year-old Lehmann, and both parties accused each other of having further damaged the reputation of Kant's last work. Thus, in 1926, Adickes finally decided to resign from the project altogether, before passing away in 1928.

Another decade passed before the *Opus Postumum* was finally published. In 1936 and 1938, the academy edition of Kant's last, unfinished work was finally published, with Buchenau and Lehmann functioning as sole editors.[8] While the conflict between Adickes on the one hand and Buchenau and Lehmann on the other was likely unrelated to political differences, the delay of the publication of the two volumes after Adickes's departure from the project in 1926 must be considered under a different light. According to biographical details provided by Christian Tilitzki in his *Die Deutsche Universitätsphilosophie in der Weimarer Republik und im Dritten Reich*, the delay of the publication might have been part of a political struggle between Eduard Spranger, the head of the Kant-commission of the *Akademie der Wissenschaften*, and the young Gerhard Lehmann, who increasingly exhibited National Socialist tendencies in both his thought and politics.[9] Given that there has been no re-edition of the *Opus Postumum*, readers are to this day reading Kant's last manuscript through Lehmann and Buchenau's edition.[10] At the same time, however, there is a conspicuous silence around Lehmann within present-day scholarship on Kant. It seems that Lehmann came to stand for the last, unfortunate chapter in the long, problematic reception history of Kant's *Opus Postumum*, in which its publication finally co-incided with National Socialism. To discuss the *Opus Postumum* thus demands a critical engagement with the political context of its publication and the writing that resulted from it.[11]

Gerhard Lehmann is important for my project beyond his role as editor of the *Opus Postumum*. Lehmann was without doubt the first thinker who has paid attention to the explicit role that technics played in Kant's philosophy. Lehmann argues together with, but substantially earlier than, Seibecke, and against Heidegger, Simondon, and Stiegler, that Kant 'knew' something about technics and took it upon himself write about it. Thus, to perform a reading of the meaning and role of technics in Kantian thought necessarily implies dealing with the Lehmann's contribution to Kantian thought.

Regarding Lehmann's explicit writing on Kant and technics, there are two main pieces of writing to take into consideration. The first is a short essay called *Die Technik der Natur* (1938),[12] and the second is Lehmann's habilitation, *Kants Nachlasswerk und die Kritik der Urteilskraft* (1939).[13] In both of these works, Lehmann argues that one should read the *Critique of the Power of Judgment* (1790) and the *Opus Postumum* (1796–1803) in continuity with one another. He also puts forward a number of claims about the role and meaning of technics in both of these works. When read in relation to each other, Lehmann proposes that these two works reveal Kant's late intention to write yet another critique of reason. Complementing the critique of pure theoretical and pure practical reason, this third critique of reason would have dealt with none other than technical reason.[14]

According to Lehmann, Kant's thought on technics was non-unified and, as such, extended across four domains (*Bereiche*).[15] The first domain, according to Lehmann, concerned itself with the idea of a technics of nature, which was first introduced in the *Critique of the Power of Judgment* before being further explored in the *Opus Postumum*. The second domain is that of technical-practical reason more generally. The third domain, according to Lehmann, is the first to have been dealt with exclusively in the manuscript of the *Opus Postumum*. Regarding this third domain, Lehmann argues that for the late Kant, the world of appearances itself is technical, as it is posited, constructed and formed by the technical acts of the subject. These technical acts are then further said to be the responsibility of the transcendental subject itself, which forms the fourth technical domain in Lehmann's view. On Lehmann's account, the *Opus Postumum* shows nothing less than that and how the transcendental subject itself is technical and engages in the technical act of self-positing.[16]

Lehmann's reading of technics in Kant, and in particular his account of the technical transcendental subject, aimed to show nothing less than that Kant's technical thought in the *Opus Postumum* extended into the depths of transcendental philosophy. Thus, while Seibicke showed that Kant introduced the shortened and Germanized term *die Technik* into German philosophy and combined the medieval scholastic notion of *Technologie* as systematization with the Aristotelian notion of *techne* as skill and ability, Lehmann further proposes that Kant's thinking of technics took place at the very heart of transcendental philosophy itself.[17]

The following chapter will predominantly work with the 1936 and 1938 *Akademie-Edition* of the *Opus Postumum* edited by Lehmann and Buchenau, with supplementary references to the online edition of the *Opus Postumum* provided by the Berlin-Brandenburgische Akademie der Wissenschaften,[18] and the Cambridge University Press English translation. I will also refer to some of the philological work that Lehmann undertook as editor of the *Opus Postumum*. Concerning Lehmann's philosophical work on the *Opus Postumum*, the subsequent chapter will follow a route in which Lehmann plays a multifaceted role. I will introduce and work with Lehmann's reading of technical-practical reason, particularly in relation to the *Opus Postumum*. Differently to Lehmann, however, who focused on the narrow relation between the *Critique of the Power of Judgment* and the *Opus Postumum*, my discussion of the role and meaning of technics for Kant's transcendental philosophy has already taken a larger frame of reference as it traces technics from the *Critique of Pure Reason* until the very last *Opus Postumum*. It is my understanding that Lehmann ultimately underestimated the technical problematic underlying Kant's critical project as a whole by presenting Kant's occupation with technics as a feature of the 'late Kant', paying no attention to Kant's notion of technics in relation to methodology, systematization and schematization. Furthermore and regarding the *Opus Postumum*, my reading of self-positing takes the transition between the first and second acts of self-positing as its key, meaning that, as the following will show, my reading of technical-practical reason develops from the middle of self-positing. The role of this transition and the role of reason as both moral-practical and technical-practical reason in this transition, I contend, have been overlooked by Lehmann. At the same time, however, and despite a number of fundamental differences between Lehmann and my own reading of the *Opus Postumum*, I will, in the following, pick up important points from Lehmann. Of particular significance here is Lehmann's precise reading of the pure forms of intuition as 'primitive technical products' of the power of imagination.

14

Technical-practical reason in the *Opus Postumum*

As the previous discussion has shown, Kant identified the problematic overstepping of theoretical reason's competences as the essential problematic that his critical method was to overcome in the *Critique of Pure Reason*. Our prior discussion of this problematic revealed not only that reason was subsequently accused by Kant of (ab)using logic as an instrument to achieve its expansions, but, further, that the twofold critical remedy of both the canon and method laid out by the *Critique of Pure Reason* harboured yet another technical remedy, as it relied on the technics of systematization and the skill of grasping by hand. The peculiarity of the ensuing discussions in both the *Critique of Pure Reason* and the *Critique of the Power of Judgment* showed in the following that Kant's discussions of technics took place within his critique of theoretical rather than practical reason, raising the question concerning the place of the technical within Kant's system. This can be seen particularly well in the *Critique of the Power of Judgment*, in which Kant developed the systematic notion of a technics of nature (*Technik der Natur*), which, while performing the difficult task of bridging the abyss between theoretical and practical philosophy, could and in fact should have nevertheless been 'appended to the theoretical part of philosophy'.[1]

The following will show that the *Opus Postumum* substantially increases the scope of technics, as well as articulating its systematic position internal to theoretical reason anew. Kant's re-elaboration of the former tripartite distinction of the imperatives of practical reason into the twofold technical-practical and moral-practical reason, the first of which is said to belong to theoretical philosophy, is thus far from being exhausted in 1797 in his *Metaphysics of Morals*. Tellingly, reason in its twofold use, features prominently in the *Opus Postum*. However, due to its status as a manuscript that was neither finished, edited, nor prepared for publication by Kant himself, and is today available to us in several, but equally challenging forms, the following pages will inquire into the *Opus Postumum* with ample philological and historical attention.

The first important mention of the term 'technical-practical reason' occurs as early as in the XIIth Fascicle, dated between February and May 1799:

> Hereby may be judged the absolute value of mathematics, in comparison with philosophy, with respect to the practical. The former is that of *technical-practical* reason (skill in the discovery of *means* for whatever ends), the latter is *moral-practical* reason and is directed to the *final end*, which is absolutely (categorically) obligatory, namely to create men of improved character [*Gesinnung*].[2]

We can here see Kant revisit his critical differentiation between, on the one hand, technical-practical reason and its technical imperatives belonging to the theory of nature, and, on the other hand, moral philosophy with the one categorical imperative proper to moral philosophy. The difference centres, once again, on the nature of the end in question. While technical-practical reason concerns the 'skill in the discovery of *means* for whatever ends [Geschicklichkeit zu Erfindung der Mittel zu beliebigen Zwecken]',[3] and thus literally the skill to invent the means to any ends whatsoever, moral-practical reason is oriented towards one single and final end, the categorical imperative, which commands categorically. This difference between technical-practical and moral-practical reason is then compared to the relationship between mathematics and philosophy. Analogous to the difference between technical-practical and moral-practical reason, mathematics and philosophy are equally distinguished regarding their ends. While both are sources of a priori cognition, mathematics does not constitute a system as an absolute whole, as a later quotation vividly affirms. Mathematics, and thus indirectly also technical-practical reason, is 'a mere product [*Kunstproduct*] of calculation and, in an indirect and only mediate (conditional) sense, directed towards ends. Philosophy, on the other hand, is directed towards absolute ends [. . .] and does not content itself with science (of means to ends)'.[4]

The second occurrence of the adjective 'technical' can be found in Fascicle X, dated by Erich Adickes to the period between August 1799 and April 1800. Here, one can find the following remarkable observation:

> The moving forces of matter are either technically or merely mechanically moving. The latter are 1) physical/mechanical and presuppose the dynamic possibility as machines.[5]

Kant proposes an explicit differentiation between the technical and the mechanical in a similar vein to the *Critique of the Power of Judgment*, where the power of imagination was said to work mechanically when under the law of the understanding in determining judgment, and freely under the technical power of judgment. And similarly to how the corresponding notion of a technics of nature was shown to be the underlying condition of possibility for the unity of nature prescribed by the mechanism of the understanding, the technical forces of nature are here said to be the condition for the physical/mechanical forces of nature.

Kant's explicit differentiation in the *Opus Postumum* between technical and mechanical moving forces is remarkable, since a large part of the *Opus Postumum* is dedicated to a continual investigation of and struggle with ether, the concept of a primordial and dynamic matter. Kant's aim is to employ this universally distributed and all-penetrating matter, referred to under the name of ether, caloric, or world-matter, to account for the fundamental characteristics of material bodies (like cohesion, rigidity and fluidity). Until around April 1799, Kant sought to deduce the elementary system of the moving forces of matter from the table of categories under the presupposition of the dynamical forces of ether. After April 1799, however, Kant can be seen to have changed his approach, as he no longer treats ether as a problem of physical phenomena. Instead, he now ventures to deduce it as a condition of possible experience.[6] Ether, Kant now claims, 'is actual, because the concept of it (with the attributes we ascribe to it) makes possible the whole of experience; it is given by reason, not as a hypothesis for perceived objects, for the purpose of *explaining* their phenomena, but rather immediately, in order to found the possibility of experience'.[7] The whole of possible experience is thus said to rest upon the condition of ether. And it is this very same ether that is called technical by Kant when he writes, 'The moving forces of matter are thus either technically or merely mechanically moving'.[8] Ether, Kant thus claims in Fascicle X, is a technical force that makes possible the whole of experience in the mechanical realm.

After ether is identified as a transcendental condition of possibility and called technical, the term 'technics', in all its grammatical forms, enters Kant's terminological arsenal with increasing frequency. From Fascicle VII onwards, written between April and December 1800, one encounters numerous references to technical-practical reason. The first mention appears in a discussion of God's activity, which is then compared to human technical-practical reason: 'Its activity is on the analogy with technical-practical reason [breaks off]'.[9] In the following pages, Kant then works through a number of formal and anthropological definitions of technical-practical reason. Technical-practical reason, we are told, aims at 'The cognition [*Erkenntnis*] of the means to any ends',[10] it is 'a relation of will towards purposes, which, with regard to himself, are unconditionally necessitating (necessitantia); if he intends to bring about this or that, then he must use this or that procedure: The imperative is conditional'.[11] Kant can thus be seen to circle again and again around what has by now become a common refrain, according to which our human will is subject to two kinds of mean-ends relation, one legislated by technical-practical reason and the other by moral-practical reason.

With the beginning of the very last Fascicle, Fascicle I, written between December 1800 and February 1803, one can see a number of interesting terminological fluctuations.[12] Kant employs the terms *technical-theoretical reason* (*technisch-theoretische Vernunft*)[13] for the first time on page 19 of Volume 22 of the academy edition, after which he returns to employing the notion of *technical-practical reason* (*technisch-praktisch Vernunft*). A few pages further along, however, he then employs the terms *technical-practical* and *technical-theoretical* in a seemingly undifferentiated manner.[14] In total, moral-practical reason is opposed to *technical-practical reason* on twenty-nine occasions in Fascicle I.[15] At the same time, however, the term 'theoretical-speculative reason' is employed as moral-practical reason's counterpart on twenty-four occasions,[16] with five mentions of the simple *speculative reason*,[17] two mentions of *technical-theoretical reason*[18] and one mention each of *theoretical-practical reason*[19] and *technical-speculative reason*.[20] Readers of the *Opus Postumum* thus find themselves in a dilemma. What kind of reason is Kant talking about here?

The vacillating terminology suggests two interrelated things. First, and most importantly, it should be noted that Fascicle I is witness to Kant's ongoing and substantial elaboration of technical-practical reason. Second, the terminological variations reveal Kant's explicit struggle to find the proper nomenclature and therefore conceptual outline for something that had seemingly hitherto escaped his systematic grasp. The fact that reason in its totality is here subjected to terminological variation and experimentation shows that Kant is indeed dealing with a systematic problem that affects reason as a whole, rather than a merely aesthetic difficulty of the adequate word for purposiveness not explicitly legislated by morality. That this is indeed a larger structural problem can also be observed in Kant's numerous experimentations with new titles for the manuscript. Strikingly, the titles sketched out in Fascicle I are a long way away from the project of a transition from metaphysics to physics, and thus the project that Kant had initially set out on. While the chronologically first Fascicle IV, written in 1796, begins with the title 'Transition from the Metaphysical Foundations of Natural Science to Physics',[21] the titles that one can find in Fascicle I now read as 'System of transcendental philosophy in Three Sections',[22] or the later 'The highest standpoint of transcendental philosophy in the system of ideas. God, the world and the human being appropriate to their duty in the world'.[23] But the title of the work in question is not the only thing that has changed drastically; so has Kant's projected table of contents. Fascicle I presents the following outline:

> Title sheet and Preface
> The world as universum
>
> In all these objects, a maximum: idea, *ergo unicum* in all three cases
>
> 1. theoretical-speculative [reason]
> 2. technical-practical [reason]
> 3. moral-practical reason
>
> From intuitions, *a priori* concepts, and ideas.
> The idea of freedom leads, through the categorical imperative, to God
>
> 1. speculative [reason]
> 2. practical [reason]
> 3. technical-practical [reason]
> 4. moral-practical reason in one system.[24]

In this new table of contents, technical-practical reason is provided with its own explicit treatment in a dedicated chapter, couched between reason in its purely theoretical and moral guises. On one side of technical-practical reason, Kant places theoretical, also called theoretical-speculative reason, concerned with nature as an object of theoretical philosophy, which the *Critique of Pure Reason* had initially attended to. On the other side, one finds moral-practical reason, and thus the subject of the *Groundwork of the Metaphysics of Morals* and the *Critique of Practical Reason*, the main object of which is freedom as the object of moral philosophy. In between these two, Kant now places technical-practical reason. Thus, while the *Critique of the Power of Judgment* had explicitly treated and supposedly secured a 'bridge'[25] between theoretical and practical philosophy by way of the power of judgment, this outline, found in the chronologically last Fascicle of the *Opus Postumum* suggests that, at the very end of his life, Kant deemed it necessary to rethink and extend the problematic of this very bridge. Indeed, it appears that his initial transition (*Übergang*) project contained in the *Opus Postumum*, to which Kant had initially given the title 'Transition from the Metaphysical Foundations of Natural Science to

Physics',[26] has not disappeared but rather diversified and multiplied as his work on the manuscript progressed.

By the time Kant arrives at Fascicle I, he now speaks of four necessary and complementary transitions: '(1) Transition from the metaphysical foundations of natural science to physics. (2) Transition from physics to transcendental philosophy. (3) Transition from transcendental philosophy to the system of nature and freedom. (4) Conclusion. Of the universal connection of the living forces of all things in reciprocal relation: God and the world.'[27] Within this cascade of necessary transitions and bridges, technical-practical reason is charged with the responsibility of providing the transition from transcendental philosophy to the system between nature and freedom. This is necessary since both nature and freedom, as Kant writes a few lines earlier in the same passage, need to be treated theoretically and practically: 'The system of knowledge which formally (thus *a priori*) precedes experience and contains the conditions of the possibility of experience in general, divides into two main branches: nature and freedom, both of which must be treated theoretically and practically; the product of technical-practical or moral-practical reason and their principles (inclination and morals/instinct-understanding) emerges.'[28] Consequently, both nature and freedom are subject to theoretical and practical treatment, and it is technical-practical reason that is now charged with supplying this twofold treatment to nature.

However, the above quotations contains an important, systematic omission. For did the *Critique of the Power of Judgment* with its idea of a technics of nature and the corresponding technical power of judgment not already perform the sought-after transition function between nature and freedom, and thus between theoretical and practical reason? Was not the idea of a technics of nature to reassure us that we are indeed at home in the world and consequently equipped with the adequate cognitive resources and faculties in order to cognize it? At the same time, the technical power of judgment was said to operate technically precisely because it compared a given natural thing to the form of a system provided by reason, whereby first opening the technical power of judgment to the question of reason's contribution to this technical operation. This question presented itself again when in the second book of the *Critique of the Power of Judgment*, discussing the power of judgment in its teleological function and having dropped any mention of the technical power of judgment, Kant now proposed that, in fact, the idea of a technics of nature designates a 'kind of causality in nature, in accordance with an analogy with our own causality in the technical use of reason [im technischen Gebrauch der Vernunft]'.[29]

Returning to the contents of the *Opus Postumum*, and in specific to the contents of the chronologically last Fascicle I, one can find no more mention of a technical power of judgment. Instead, it is now technical-practical reason, that is, reason in its technical use, which is charged by Kant with the responsibility of treating nature technically-practically in order to 'complete' theoretical philosophy. The question from hereon is, then, how exactly technical-practical reason's twofold contribution to theoretical reason is to be thought.

On the relation between Fascicles VII and I

The preceding section has already mentioned that a more systematic employment of the term 'technical-practical reason' first arises in the contents of the chronologically second-from-last Fascicle VII, which, as a whole, is dedicated to Kant's discussion of positing (setzen) and self-positing (selbst-setzen). As Frederick Beiser pointed out, the term *Selbstsetzungslehre* (doctrine of self-positing), which is how the contents of Fascicle VII are often referred to within secondary literature, is a scholarly anachronism.[30] Kant himself never used the term in the substantive grammatical form, while he did frequently refer to the substantive self-affection (*Selbstaffektion*). Kant's engagement with technical-practical reason thus arises within the context of a discussion on positing and self-positing. However, his engagement with technical-practical reason is not limited to Fascicle VII but continues in the following and final Fascicle I. Many interpretations of self-positing in the *Opus Postumum* however limit their discussion of self-positing to Fascicle VII, with references to the preceding Fascicles X and XI and their discussion of self-affection, as for instance in the work of Eckhart Förster, Dina Emundts and Bryan Hall.[31] After all, Kant started a whole new Fascicle, Fascicle I, in which he turned to address the topic of reason and its ideas, including an extensive experimentation with new titles for his work. It thus might appear that Fascicle I contained sketches for a second, altogether different book.[32] Furthermore, Fascicle I is certainly a difficult piece of writing. Differently to the other Fascicles, the wrapper of Fascicle I is entirely covered with notes and comments, due to which the academy edition does not distinguish it from the main body of the text anymore. Within the Fascicle itself, one finds frequent biographical references to Kant's age, his bodily state, as well as the circumstances of his life with increased frequency as work on the manuscript became increasingly difficult according to Kant's increasing senility, according to Adickes. This is most visible, Adickes argues, because over more than two years, Kant wrote only about 120 print-pages worth of material, which is substantially less than his 'normal', supposedly pre-senile daily pensum.[33] At the same time, however, and as the following will show, when Kant does manage to bring his thoughts to paper, one encounters a lucid mind, which aims to expand self-positing to reason and its ideas, as well as develop a plan to a full system of, rather than the mere conditions of possibility for, transcendental philosophy.

Consequently, it is my understanding that the discussion of Fascicle I, which focuses on reason and its ideas, does not stand in an external relation to Fascicle VII. Fascicle VII consists of several *Beylagen* (supplements). *Beylagen I-VIII* are signed by Kant himself. *Beylage V* is, however, what Vittorio Mathieu called a 'Pseudo-Beylage [pseudo-supplement]'.[34] While placed at the centre of Fascicle VII in the academy edition,[35] it is in fact written on the back of a letter written by Wasianski to Kant on 19 December 1801, meaning that it chronologically falls within the writing period of Fascicle I.[36] What appears to have happened is that the originary *Beylage V* must have slipped into the contents of Fascicle X (Fascicle X, sheets 59–61), consequently, we can find a note on *Beylage VI*, that 'perhaps it should have been *Beylage V*' which, according to Matthieu, confirms his theory of the loss of the originary *Beylage V*. It remains unknown as to why Kant proceeded to call a section from Fascicle I *Beylage V*.[37] Matthieu takes this as an indication that Kant has, in fact, intended to relate the problem of God and the world, and thus the topics of the final Fascicle I to the issues treated in the *Beylagen*, that is, self-positing.[38] The subsequent *Beylagen IX* and *X* do not contain the proper heading of a *Beylage* anymore. In the academy edition, one finds a roman '*IX*' written by Kant's hand himself, while someone else has written '7. Convolut Bogen IX'.[39] The same counts for *Beylage X*. Also here, a different hand has written '7. Convolut Bogen X' on the very top of the page. The Fascicle itself carries the simple heading 'Ich bin [I am]', written by Kant himself.

These so-called *Beylagen IX* and *X* are generally considered to contain the transition from Fascicle VII to the chronologically last Fascicle I. Hence, it comes as no surprise that some of the language as well as topics that Kant discusses in Fascicle I find themselves anticipated on the last pages of Fascicle VII. The same, however, also applies to *Beilage V*. While technical-practical reason is mentioned a handful of times in *Beylagen IX* and *X*, it is in *Beilage V* that one encounters self-positing as discussed from the perspective of technical-practical and moral-practical reason.

It is understandable that one might hesitate to read reason, the main subject of Fascicle I, and self-positing, the main subject of Fascicle VII, through one another. From the perspective of the *Critique of Pure Reason*, the faculty of reason as portrayed in the *Opus Postumum* transgresses the critical limits set by the understanding as it comes onto the stage and assumes an active role and participation within the discussion on self-positing, and thus within Kant's re-elaboration of self-consciousness and the transcendental apperception. However, I believe it nevertheless to be necessary to read Fascicles VII and I in continuity with one another due to the conceptual discussion of technical-practical reason itself, as its function and role, as the following will show, is essentially within and for the process of self-positing. As such, I follow Matthieu in his proposal to read Kant's Fascicle I in relation to self-positing and thus the contents of Fascicle VII.[40] What, then, is technical-practical reason's contribution to and place within self-positing?

15

Technical-practical reason and the doctrine of self-positing

As noted already, Kant's deliberations on technical-practical reason systematically set in with the transition from Fascicle VII to I. The explicit topic of Fascicle VII itself is self-positing (*selbstsetzen*), meaning that Kant's engagement with technical-practical reason emerged within his discussion of self-positing, and thus the process by which the subject makes itself into an object of experience. The following chapter contains a discussion of self-positing, in order to then, in a second step, locate the role and meaning that technical-practical reason plays within it. I will closely follow Kant's account of self-positing, which in most places is explicated as two main acts.[1] Before moving on to the second act in the following chapter, the present chapter will focus on the first act of self-positing and, more importantly, the transition between the first and second act, as I will argue that it is in the very transition between these two acts that reason, as both technical-practical and moral-practical reason, enters the discussion. In order to appreciate the full contribution of technical-practical reason to self-positing, my argument will focus on the following three points. First, I will present an account of Kant's renewed engagement with the doctrine of ideas at the centre of the transition. Second, then, I will specifically focus on the idea of world, and Kant's discussion of knowledge of the world (*Weltkenntnis*) and philosophy in the world concept, both of which will bring us back to essentially methodological questions contained within the problem of transition. Third, I will argue that these transcendental, methodological discussions take on a function previously discussed under the name of transcendental reflection. Following the discussion of these three points, I will then, in the final section of this chapter, move on to the second step of self-positing, which, I will argue, consists of the twofold moral-practical and technical-practical modes of self-positing.

In *Kant's Opus Postumum dargestellt und beurteilt* (1920), Erich Adickes characterized Kant's reflections on positing (*setzen*) as a quasi-response and admission to absolute idealism. Adickes claim is that after students and readers of Kantian philosophy, such as Beck, Reinhold, Abicht, Maimon and in specific Fichte set out in search of a highest principle of transcendental philosophy – with each of these thinkers aiming to deduce the elements of our faculties from a single unifying and supreme principle – Kant embarked on such a search himself in his final project.[2] Concerning Kant's relation to Fichte, Kant had been an avid reader of Fichte's early writings, and the two had exchanged letters since 1791.[3] The 1794 publication of Fichte's *Wissenschaftslehre*, however, caused Kant to publicly distance himself from Fichte, denouncing the *Wissenschaftslehre* as 'a completely untenable system'.[4] Kant's own explicit engagement with positing in the *Opus Postumum* does indeed begin after his public critique of Fichte, with positing first appearing in Fascicle X and XI. The contents of Fascicles X and XI were written from 1799 onwards, suggesting that Kant might have written his own doctrine of self-positing with and against Fichte.

Eckhart Förster is, however, right to insist that Kant had already engaged with positing as early as in his 1763 *The Only Possible Argument in Support of a Demonstration of the Existence of God*.[5] It was in this early text that Kant first proposed that existence is not a real predicate of a thing and that 'the concept of position or positing [*setzen*] is completely simple and identical with the concept of being in general'.[6] Here, Kant differentiates between two modes of position, on the one hand a relative position and on the other an absolute position. In a relative position, a predicate is related to a subject in a judgment via a copula, as in 'S is P'. In such a judgment, what is posited is a property or predicate of the subject. Consequently, Kant explains, relative positions are identical to 'the copulative concept in a judgment',[7] the relative position posits a predicate in relation to a subject. When it comes to an absolute position, however, it is the subject itself that is posited, as in 'S is'. Existence, according to Kant, is thus an absolute position.

Förster is not the only one who uttered doubts about attempts of relating Kant's engagement with self-positing as his response to Fichte *tout court*. According to Adickes, Kant supposedly lost the ability to think himself into lines of thought foreign to his own from as early as the 1790s.[8] Adickes argues by way of letters and other biographical material that – fully entrenched in working out his own system of transcendental philosophy – Kant never studied Fichte's system or any of his other contemporaries. At the same time, he nevertheless admits of the possibility that Kant might have learnt about the general outline and method of Fichte's theory of self-positing from conversations, reviews, as well as hear's say.

While Kant's relation to and knowledge of Fichte's search for a highest principle is essentially difficult to pin down philologically and historically, the problem of which we will return to in the final chapter, Kant's relationship and insight into Beck's theory of originary representation can be substantiated by both direct references in the *Opus Postumum* as well as from letter exchanges and conversations between the two thinkers. In 1793 and 1794, Beck had famously published his *Erläuternden Auszug aus den kritischen Schriften des Herrn Prof Kant*,[9] an explanatory, insightful compendium to Kant's critical philosophy. While the first two editions are dedicated towards explication and explanation of transcendental philosophy and, as such, do not traverse the limits of Kantian philosophy themselves, the third edition, published in 1796, is an altogether different case. It's subtitle *Einzig-möglicher Standpunkt, aus welchem die kritische Philosophie beurteilt werden muß* already points to Beck's intention of providing a single unifying principle for transcendental philosophy, by way of which the elements of our faculties could then lawfully be deduced in their totality. Against Reinhold, Abicht, Maimon and the others, Beck argues that the beginning of transcendental philosophy must not be a principle that itself consists of concepts, but is rather an issue of an originary postulate, like that of a geometer, according to which an object can originarily be represented.[10] A number of direct references to Beck's originary representing can be found throughout the *Opus Postumum*, as for instance in Fascicle X, when Kant writes, 'That we have insight into nothing except what we can make ourselves. First however, we must make ourselves. Beck's original representing'.[11]

Eckhart Förster traces Kant's engagement with self-positing back to a note,[12] written between 1788 and 1790 and thus to Kant's ongoing discussions with J. G. C. Kiesewetter. In this note, Kant writes that 'I posit my own existence [. . .] for the sake of empirical consciousness and its possibility [. . .] myself as a being that exists in a world'.[13] Kant can here be seen to directly refer to the amendment undertaken in the 1787 B-Edition of the *Critique of Pure Reason*, where the representation of myself as a subject of thought ('I think') was shown to be itself dependent on something empirical being given to it, since, without it, thought and combination would have nothing to exercise its spontaneity on. Consequently, Kant claims, that 'the proposition "I think," insofar as it says that **I exist thinking**, is not a merely logical function, but rather determines the subject (which is then at the same time object) in regard to existence'.[14] Importantly, then, the determination of my consciousness in time essentially presupposes that I exist and occupy a position in space, to which other things can, therefore, be external to.

How exactly such a position of externality can be thought of remained, unresolved within the context of the *Critique of Pure Reason* according to Förster, since space, for the transcendental idealist point of view, is the mere form of outer intuition rather than designating something actually 'outside' myself. Förster's point was consequently that in order to renegotiate the relation between the 'I think' of the spontaneous thinking subject and myself as a spatially extended, corporeal and existing being in space, self-positing, as the act by which the subject posits itself in the sense of making itself into an object of experience, needs to address both the problematic of space and my position in it.[15] Thus, as concerns the relation between the *Opus Postumum* and the *Critique of Pure Reason*, one can broadly say that self-positing deals with the problem of synthesis, the copulative concept in a judgment, and thereby, in a second step, with the relation between the transcendental apperception of the 'I think' and the empirical consciousness 'I am', a subject matter which had already been the explicit subject of the *Transcendental Deduction* in the *Critique of Pure Reason*.

Self-positing in the *Opus Postumum*

Given that self-positing is thus a topic that had already long occupied Kant, he begins his re-investigation of self-positing within the context of the *Opus Postumum* in Fascicle X, drafted from August 1799 to April 1800. The first explicit mention of self-positing is the already mentioned reference to Beck, according to which 'we have insight into nothing except what we can make ourselves. First, however, we must make ourselves. Beck's original representing'.[16] This reference engages two moments of the *Critique of Pure Reason*. On the one hand, it recalls the B-Edition of the *Transcendental Deduction*, where Kant proposed that 'I exist as an intelligence that is merely conscious of its faculty for combination'.[17] On the other hand, it further refers to Kant's so-called 'Copernican turn', according to which 'we can cognize of things *a priori* only what we ourselves have put into them [*in sie legen*]'.[18] Consequently, the *Opus Postumum* does not propose a perspective that had been absent in the *Critique of Pure Reason*. The *Critique of Pure Reason* had already claimed that self-consciousness means being conscious of one's faculty of combining, and, furthermore, that objects conform to our knowledge rather than the other way around.

The quotation from the *Opus Postumum* does, however, seem to contain an intensification, in the sense that 'what I can combine' is reformulated into the more general 'what I can make', and the previous 'consciousness is consciousness of my faculty of combination' becomes 'consciousness is consciousness of my ability to make myself'. This means that the terminology of *machen* (to make) and *können* (to be able to), and thus one's acts, skills or abilities, has taken over from the previous focus on combination more broadly that Kant had employed throughout the *Critique of Pure Reason*. In the *Opus Postumum*, the conditions of possibility thus appear to have become conditions of what we can make.[19] And in order to be able to delineate that which we 'can make', we first need to 'make ourselves', in the sense of positing oneself as existing. How is it, then, that we 'make ourselves'?

There are a multitude of terms employed by Kant that designate the act of positing. Alongside *setzen* (to posit), Kant also employs *machen* (to make), *vorhermachen* (to pre-make), *constituieren* (to constitute), *zum Object machen* (to make into an object), as well as *sein Selbstbewußtsein determinieren* (to determine one's self-consciousness) and others.[20] The Latin term *positus* itself, the participle of the verb *ponere*,[21] means to place, to put, to posit, to put up, to present and lastly, to bury. In Fascicle VII, Kant offers a systematic breakdown of the different elements and moments that together make up self-positing. First, Kant differentiates between our self-consciousness as *facultas representativa* and 'the determination of oneself as function of oneself as *vis representativa*'.[22] Second, in critical fashion, he further differentiates between the subject in appearance and the subject as a thing in itself, objectively=x. This means that with regard to positing, we have to attend to four interconnected, but differentiated, elements: first, the *facultas representativa* (faculty of representation); second, the *vis representativa* (power of representation); third, the subject as appearance; and fourth, the subject as a thing in itself.[23] These different elements are produced by acts (*Akte*), which in most places are said to be two in number.[24]

The first act of self-positing

'The first act of the faculty of representation (*facultas repraesentativa*) is the representation of oneself (*apperceptio*) through which the subject makes itself into an object (*apperceptio simplex*)'.[25] Apperception is here said to be originary in two interconnected ways. First, 'this act of consciousness (*apperceptio*) does not arise as a consciousness of something preceding'.[26] Consciousness of myself as an object of thought is thus defined as a spontaneous and originary act, it is not triggered or caused by anything preceding it. As such, the consciousness of myself as an object cannot be derived by syllogistic reasoning. Similarly to the *Critique of Pure Reason*, one finds numerous instances in the *Opus Postumum* where Kant continues to distance himself from the conception of the Cartesian *cogito*. Second, this first act of self-positing is said to be an act of self-differentiation, in the sense that the faculty of representation 'goes from within outward with something that it itself posits'.[27] Importantly, however, this first act is merely logical and analytical. As such it is a merely formal undertaking and contains no content or material at this point.[28] That which is posited is 'the formal aspect of judging according to the rule of identity'.[29] Proceeding according to the formal rules of identity and non-contradiction, the first act of self-positing is thus an act of pure and mere thought, which concerns itself with our mode of cognition in general and the object of which is nothing but the faculty of representation itself. Kant's point here is that the formal conditions of judgment itself, the 'subject-copula-predicate' form, are the result of something that is done. They are made possible by an originary act rather than being 'native' or immanent to our faculty of representation as a pre-requisite for judgment. This originary act, in its first moment, is an act of separation and differentiation, in which one reflexively inquires into (*erkunden*) and measures (*ausmessen*) one's own faculty of representation.[30] In so doing, one differentiates oneself into a subject (I, the one who acts) and an object (I, the one who is being inquired into, my faculty of representation). Relating these two purely formal terms to one another is the copula *sum*,[31] Latin for 'I am', which, equally to myself as a predicate, is being pulled out of 'I'. The first act of self-positing thus results in the differentiation of myself into a merely formal subject and object, both of which were already identically contained in the subject, albeit in a less distinct form.

This discussion so far coincides with Kant's claims in the *Critique of Pure Reason*, where the transcendental apperception was specified as consciousness 'of myself not as I appear to myself, nor as I am in myself, but only that I am. This representation is a thinking, not an intuition'.[32] But the B-Edition of the *Critique of Pure Reason* has also argued that the thought 'I am' essentially presupposes something given, since without it there would be nothing to exercise my spontaneous thought upon. And rather than only the apperception of myself as an object of thought, it is the very relation between myself as an object of thought and an object in space and time that is, properly speaking, the problem of self-positing.

From the first to the second act of self-positing

In order to address this relation between myself as an object of thought and an object in space and time, we must move on from the first to the second act of self-positing. While the first act was said to be the responsibility of the faculty of representation (*facultas representativa, Vorstellungsvermögen*), the second act is the doing of the power of representation (*vis representativa, Vorstellungskraft*), which is in general charged with the responsibility of synthetically combining intuitions and concepts. It is thus in this second act that the merely logical form of judgment is to be supplied with content (intuition) that will allow the merely logical object that I am to become an object (*Gegenstand*) in the proper sense.

There is, however, a transition between the first and the second act of self-positing, which Kant calls a progression (*Fortschritt*): 'Only by the subject becoming its own object does apperception proceed [*schreiten*] to apprehension, and proceed from metaphysics to transcendental philosophy, from the analytic to the synthetic a priori.[33] This progression between the first and the second act of self-positing thus passes from metaphysics to transcendental philosophy, and from analytic to synthetic judgment. How, then, is this progression conceptualized?

The terms that Kant employs to describe the transition or progression from the first to the second act of self-positing are *schreiten* (to step or progress),[34] *hervorgehen* (to emerge),[35] *führen* (to lead)[36] and *sich reihen* (to follow in a series).[37] The second act of self-positing is thus said to emerge out of the first, the first is said to 'lead to' to the second act, 'logical consciousness [progresses] to what is real',[38] *es reihet sich* (it constitutes a series).[39] Thus a second act emerges out of the first act, and this second act must provide the empty and merely logical form of the object that I am with content. This means that content (intuition) is yet to be introduced into the equation, and that, in a consecutive moment, these heterogeneous elements of intuitions and concepts are then to be combined and applied to one another.

In the *Critique of Pure Reason*, the mutual preparation and combination of intuitions and concepts was the explicit responsibility of the power of judgment (*Urteilskraft*). When it comes to the *Opus Postumum*, the second act of self-positing is equally said to be the responsibility of the power of judgment, the forms of which are said to emerge analytically from the faculty of representation, meaning that they are to be pulled out of the faculty of representation in another judgment of clarification.[40] However, Kant proposes that next to the forms of the power of judgment there is a second element, which also emerges in the transition from the first to the second act of self-positing. Here Kant introduces a radical difference to the *Critique of Pure Reason*: 'The principles of the possibility of making [*anzustellen*] experience lie identically in the faculty of representation in general (facultas repraesentativa), from which the forms of the power of judgment emerge analytically and the principles of reason apodictically'.[41] The transition from the first to the second act is thus said to also be the place where the principles of reason apodictically, meaning necessarily, come onto the stage.

Fascicle I repeats that the place of reason and its ideas must be somewhere in the progression from the first to the second act of self-positing. Ideas, Kant writes here, come onto the scene 'still prior to a self-determination in space and time (although, for the purpose of it)'.[42] This means that 'out of' the purely logical consciousness *I am*, and before any further act of determination of the object that I am as an object of possible experience in time and space, both reason and the power of judgment emerge. And while the *Critique of Pure Reason* claimed that the power of judgment is the responsibility of the *Analytic of Principles*, which famously contains the principles 'without which no object can be thought',[43] including myself as an object in time and space, it is yet to be seen why reason needs to enter in the transition from the first to the second act of self-positing.

Ideas, ideals, original images

Subsequently, then, reason, within the *Opus Postumum*, is said to 'make' ideas, just like the faculty of representation in the first act of self-positing makes itself into a formal object of thought. Ideas, Kant writes, are self-made (*selbstgeschaffen*).[44] Reason thus comes onto the scene between the first and second act of self-positing explicitly as a source of creation and invention, as the originator of ideas, meaning that ideas originate autonomously and spontaneously within reason rather than following, as a reaction, from any form of outward affection.[45] As such, the *Opus Postumum*, however, deviates further from the path laid out in the *Critique or Pure Reason*, since there reason was hitherto denied any role as a creative, generative force: in the *Antinomy of Pure Reason* Kant had famously claimed that 'reason really cannot generate any concept at all, but can at most only free a concept of the understanding from the unavoidable limitations of possible experience, and thus seek to extend it beyond the boundaries of the empirical, though still in connection with it'.[46] Consequently, while in the *Critique of Pure Reason* the understanding was affirmed as a generative faculty, the faculty of reason was at the same time denied a productive function, as ideas themselves were specified as 'nothing except categories extended to the unconditioned'.[47] It is thus in clear distinction to the *Critique of Pure Reason* when, in the *Opus Postumum*, ideas are now explicitly subject to being made by reason as they enter the scenario of self-constitution prior to the second act of self-positing. Since ultimately the horizon of this discussion concerns the role and meaning of technical-practical reason for self-positing, what remains to be shown in the following is whether reason in its entirety is productive of ideas, meaning that technical-practical reason 'makes', or whether such transcendental 'making' is restricted to moral-practical reason only.

Reason, Kant writes, 'goes ahead with the sketching [*Entwerfung*] of its forms'.[48] And in the last Fascicle I he repeats that 'Ideas precede appearances in space and time',[49] affirming once again the placement of reason prior to any further act of self-determination. At the same time, Kant still holds that, as concepts of reason, ideas are thought-entities and thus 'thought entities, subjective products of its own human reason, which the subject relates to itself'.[50] Thus, not only do ideas originate within our own human reason, they are, second, restricted to the realm of thought (*Gedankenwesen*) and are, third, reflexively directed only towards ourselves. What being directed towards itself means is that they effectively function as 'laws of thought which the subject prescribes to itself. *Autonomy*'.[51] Ideas thus perform the role of immanent self-legislation. But differently to Kant's position in the *Critique of Pure Reason*, ideas are here said to organize our system of thought and legislate prior to having received any content or object, or even developed self-consciousness in the full sense. This means that there must be an essential relation between reason's ideas and the second act of self-positing in the sense that reason's ideas essentially contribute to and work towards the second act of self-positing.

While the *Critique of Pure Reason* had explicated ideas of reason as discursive in nature and thus as belonging to the conceptual realm, the *Opus Postumum* now claims that finally, 'ideas are not concepts, but pure intuitions: not discursive, but intuitive representations, for there is only one such object'.[52] This last point on the status of ideas as intuitions is repeated in slightly different iterations throughout the manuscript: 'Ideas are images [*Bilder*] (intuitions), created *a priori* through pure reason, which, [as] merely subjective thought-objects and elements of knowledge precede knowledge of things. They are the archetypes (*prototypa*)'.[53] Thus, while in the *Critique of Pure Reason* Kant positioned his own thinking of ideas somewhere mid-way between Aristotle, who restrained the idea to the empirical, and Plato, who considered the idea to be a form of archetype,[54] the *Opus Postumum* now suggests that Kant's conception has moved a considerable step closer to the Platonic archetype and thus towards what in the *Critique of Pure Reason* and the *Critique of the Power of Judgment* was called an ideal.

In the *Critique of Pure Reason*, an ideal was defined as 'the idea not merely *in concreto* but *in individuo*, i.e., as an individual thing which is determinable, or even determined, through the idea alone'.[55] As such, ideals are archetypes, with the Stoic sage and God functioning as examples.[56] According to the subsequent *Critique of the Power of Judgment*, ideals are

> the representation of an individual being as adequate to an idea. Hence the archetype of taste, which indeed rests on reason's indeterminate idea of a maximum, but cannot be represented through concepts, but only in an individual presentation, would better be called the ideal of the beautiful, something that we strive to produce in ourselves even if we are not in possession of it.[57]

As such, in the production of beautiful art the products of genius are said to be singular examples of the ideal of beauty, and thus to function as exemplary archetypes for further production.

When it comes to the *Opus Postumum*, Kant now claims that ideas are presented in intuition as ideals, or *Urbilder* (original images).[58] Each one of these original images is, first, numerically singular, 'each of these objects is absolutely one (*unicum*)'.[59] Second, each is a maximum. In continuity with the *Critique of Pure Reason*, Kant proposes that the world, a quantitative maximum, is one of reason's ideas, while the idea of God is its qualitatively maximum counterpart.[60] Third, both of these ideas are 'infinite; the first as magnitude of appearance in space and time, the second in terms of degree (virtualiter)'[61] The world is the idea of an infinite, singular maximum in the phenomenal realm of space and time, while the idea of God is its noumenal counterpart. Thus, there is, on the one hand, God, who is said to have only rights but no duties and who, as it were, unites all objects of pure reason. On the other hand, there is the idea of world, which concerns all empirical objects and effects their unification.

While the concept of ether does surface occasionally during the very last Fascicle I, Eckhart Förster is in an important manner right to point out that the idea of world must essentially be understood as 'the successor to the ether conception of the earlier fascicles'.[62] Similarly to how Kant had arrived at a conception of ether 'given by reason, not as a hypothesis for perceived objects, for the purpose of *explaining* their phenomena, but rather immediately, in order to found the possibility of experience itself',[63] the idea of world (and all self-made ideas in general) are explicitly said to not be 'hypothetical (problematic or assertoric) but apodictic, as it [reason] creates itself'.[64] Furthermore, in as much as all objects are intuited within the absolute whole called God, Kant asserts that 'likewise, we can say that they [all objects] must, regarding their reality, be encountered in the world'.[65] This is the case because the essential difference between the objects intuited in God and the objects intuited in the world does not lie within the objects themselves but is rather to be located 'in the difference of the relation in which the subject apprehending the sense object is affected for the production of the representation in itself'.[66]

God, the world and the human

Complementing reason's original images of God and the world, Kant places a copula, meaning a further mediating concept capable of reconciling the heterogeneous wholes of God and the world. Consequently, it becomes clear that reason's self-made ideas are internally differentiated, in the sense that they are characterized by uneven relations to one another:

> God and the world are two objects of transcendental philosophy; thinking *man* is the subject, predicate and copula. The subject who combines them in one proposition. These are logical relations in a proposition, not dealing with the existence of objects, but merely bringing what is formal in their relations of these objects to synthetic unity: God, the world, and I, a man, a world-being myself, who combines them.[67]
>
> *God*, the *world*, and *man* as a person: that is, as a being who unites these concepts.[68]
>
> God, the world, and what unites both into a system: the thinking, innate principle of man in the world (*mens*). Man as a being in the world, self-limited through nature and duty.[69]

Concerning the doctrine of ideas, there is, then, another fundamental change undertaken in the *Opus Postumum*, since complementing God and the world, the *Critique of Pure Reason* had hitherto placed the soul. Now, however, the ideas of God and the world are both complemented and combined by the idea of the human.

Looking back to the *Critique of Pure Reason*, Kant had argued that ideas 'are not arbitrarily invented, but given as problems by the nature of reason itself, and hence they relate necessarily to the entire use of the understanding'.[70] These problems were then said to be threefold in nature, meaning that there are three different classes of ideas, which are derived from the three categories of relation when extended to the unconditioned. The first idea 'contains the absolute (unconditioned) unity of the thinking subject, the second the absolute unity of the series of conditions of appearance, the third the absolute unity of the condition of all objects of thought in general'.[71] This means that the first idea contains the relation to the thinking subject, the second to the world as the sum of all appearances, and finally the third to all things in general. From these three classes of ideas, Kant then deduced the three sciences of special metaphysics – Rational Psychology, Cosmology and Theology – which are treated one after another in the *Transcendental Dialectic of Pure Reason*. Kant had inherited this threefold structure of special metaphysics (Rational Psychology, Cosmology and Theology) from Leibniz and Wolff. Rational Psychology was famously given privilege within the overall structure of the *Critique of Pure Reason*, in the sense that the entire *Transcendental Analytic* must be understood as referring to the operations of a transcendental subject, which was then restricted and further qualified in the *Transcendental Dialectic*.

When it comes to reason's ideas within the *Opus Postumum*, Kant now carries out a fundamental change to the *metaphysica specialis*. Reason is still said to make three ideas. On the one hand, there are still God and the world as the objects of Theology and Cosmology, respectively. The third idea, however, has now become the object of Anthropology rather than Rational Psychology, as it concerns the human rather than the soul.

The growing importance of anthropology within and for Kant's transcendental philosophy had already been announced earlier. Kant famously added a fourth question to what the *Critique of Pure Reason* had articulated as reason's threefold interest in the publication of his lectures on *Logic* (1800). While the *Critique of Pure Reason* stated only a threefold interest of reason with the questions *What can I know?*, *What can I do?*, and *What may I hope?*, the *Logic* added the question of *What is the human?*[72] This fourth question must not be misunderstood as merely an external addition. Rather, the first three questions are said to be related to and are in fact aimed at the fourth question. In essence, 'we could reckon all of this as anthropology, because the first three questions relate to the last one'.[73]

This important change to the critical programme was first raised by the German philosopher Max Scheler in *Die Stellung des Menschen im Cosmos* (1928), when he argued that every speculative philosophy, including the one put forward by Kant, must eventually ask the question concerning the essence of the human. The importance given to anthropology by the late Kant has shaped the course of twentieth-century post-Kantian philosophy dramatically, leading to a bifurcation into, on the one hand, a philosophy aiming to found itself on anthropology and, on the other, a philosophy aiming to overcome this problematic, anthropological dimension. However, and as pointed out by Christian Bermes in his *Welt: Vom metaphysischen zum natürlichen Weltbegriff* (2004), one would do well to question whether, in fact, Kant already had in mind the same notion of anthropology that would later be developed by Scheler, Plessner, Gehlen, as well as Heidegger.[74] According to Bermes, what is rather important is that Kant both introduces and explicates the fourth question within the context of his discussion of philosophy as a world concept (*Weltbegriff*), a point that one also finds in Foucault's *Introduction to Kant's Anthropology*, containing Foucault's proposition that when it comes to the *Opus Postumum*, 'all reflection on man involves reflection on the world".[75]

World and knowledge of the world

The world is one of reason's three ideas, and as such, addressed in the *Transcendental Dialectic* of the *Critique of Pure Reason*. What this means is that, according to the specification of ideas versus concepts of the understanding, the world can neither be known in the sense of being determinately cognized nor is it an ontological presupposition. Instead, the world essentially only 'exists' in thought, it is an idea of reason. Reason employs the idea of world to situate appearances, which are under the rule of the categories, in relation to one another. Thus, as Eugen Fink pointed out in his 1985 *Einleitung in die Philosophie*, Kant's intervention into cosmology in the *Critique of Pure Reason* effected that the world was, for the first time, no longer considered and accounted for as a 'being, not a thing [. . .], that any attempt to think of it as a thing [. . .] entangles human reason in an insoluble contradiction'.[76] In its status as an idea, the world is thus not a something that can be given (*gegeben*), but it is rather *aufgegeben* (a task), 'more [. . .] a republic to be built than a cosmos given in advance'.[77]

The *Critique of Pure Reason* thus already problematized cosmology in its proposition that the world is both a problem and a task (*Aufgabe*). Hegel and other German idealists would continue to develop what Kant started under the name of the world, as a problem of speculation and the system.[78] Differently to Hegel, for whom the movement of at once the world and reason would raise the important question of history, Kant addresses the problem of the world by way of two complementary forms of knowledge. On the one hand, the world as a task is the responsibility of scientific knowledge proper. Second, however, it is also the responsibility of another form of knowledge that Kant calls *Weltkenntnis* (knowledge of the world), which he specifies as a propaedeutic and pre-scientific *Kennen* (knowledge) of the world.[79] This notion of knowledge of the world (*Weltkenntnis*) must not be understood as separating off and treating an altogether different domain of being, as it is merely a methodologically and epistemologically different approach to the world than the one provided by science. How, then, is one to understand this propaedeutic, pre-scientific knowledge of the world?

Kant taught seminars on knowledge of the world (*Weltkenntnis*) throughout his entire career, having on numerous occasions been vocal in underlining the necessity of providing his students with the essential 'knowledge of historical matters which could make good their lack of experience'.[80] Initially, these seminars were held under the all-encompassing title of geography. In the year 1765, Kant decided to divide the encompassing seminar on geography into the three disciplines of physical, moral, and political geography. The second, moral part of geography concerned the human, and from 1772 onwards, it came to form a stand-alone lecture course on anthropology before the collection of lecture notes were published in book form in 1798 under the title *Anthropology from a Pragmatic Point of View*. It was not until 1802 that the longer and more encompassing publication derived from Kant's lectures on *Physical Geography*, edited by Friedrich T. Rink, finally followed.[81]

The introductions to both of these lecture courses provide important insight into the nature of knowledge of the world. There is, Kant writes in the Introduction to *Physical Geography*, 'a great need for instruction in how to apply one's knowledge and make use of it in a manner appropriate to one's understanding and present situation, or to provide a practical use for one's knowledge. This constitutes knowledge of the world'.[82] Knowledge of the world thus addresses the problem of application and use of theoretical knowledge in the world: 'The world is the foundation [*Substrat*] and stage [*Schauplatz*] on which our ingenious play [*Spiel unserer Geschicklichkeit*] is performed. It is the ground on which we obtain and apply our knowledge.[83] The world is here explicated as the stage and thus ground for theoretical knowledge, and it is only by being situated in relation to this stage that theoretical knowledge is given both its meaning and purpose.

While itself a natural form of knowledge, pre-scientific and pre-supposable in every human, the notion of world at work in this form of knowledge is nevertheless itself systematic and architectonic. We have to, Kant writes, 'become acquainted with the objects of our experience as a whole. Thereby, our knowledge is not an aggregation but a system'.[84] As a propaedeutic, the study of anthropology and physical geography thus aims to cultivate the 'idea for knowledge of the world. What we are doing here is making an architectonic concept for ourselves, which is a concept whereby the manifold parts are derived from the whole',[85] rather than the whole being assembled from its parts. As such, it anticipates

> our future experience in the world, giving us, as it were, a pre-formed conception [*Vorbegriff*] of everything. We say of someone who has travelled widely that he has seen the world. But knowledge of the world is more than merely seeing it. Anyone who wants to derive benefit from a journey must make a plan in advance, and not regard the world merely as an object of the outer sense.[86]

Knowledge of the world is thus at once a problem of the application of theoretical knowledge and of the anticipation of experience by means of an architectonic idea of a whole – the world – in us.

As a system, knowledge of the world is explicitly the combined effort of both physical geography and anthropology, and thus concerns both nature and human beings. 'Both parts, however, must be considered *cosmologically*, namely not with regard for a single noteworthy object that they contain (physics and empirical psychology), but rather with regard to the relationship to the whole in which they are found, and within each takes its place.[87] It is thus not only that anthropology and geography belong together as sister-disciplines, but moreover that they must themselves be considered cosmologically, and thus in respect to the whole of which they are each a part.[88] The world is a cosmological whole, 'that has already been acquired as a specific, natural experience and can also be cultivated in this form'.[89] Thus, when Kant, in his lectures on logic, adds the fourth question regarding the human to the hitherto threefold programme of critical philosophy, he cannot be said to 'reduce' critical philosophy to the human. He rather grounds and relates critical philosophy to a larger notion of the world, of which both geography and anthropology are a part.[90] Kant's point, we can thus conclude, is that the three questions of critical philosophy need to be addressed in relation to their 'stage', that is, the world.

Philosophy in the world concept

In 1781, the *Critique of Pure Reason* had already differentiated between two different kinds of knowledge when it comes to philosophy.[91] On the one hand, and until now, Kant writes that 'the concept of philosophy has been only a scholastic concept [*Schulbegriff*], namely that of a system of cognition that is sought only as a science without having as its end anything more than the systematic unity of this knowledge, thus the logical perfection of cognition'.[92] Distinct from this scholastic concept, Kant places what he calls 'philosophy *in sensu cosmico*'[93] and thus philosophy in the world concept, which Alfredo

Ferrarin also calls cosmic philosophy.[94] What ultimately differentiates the two kinds of philosophy according to Kant is their form of presentation and their end, meaning that their difference is to be found in their methodology rather than concerning a difference of domain. Kant's critique of scholastic philosophy argues that scholastic philosophy is neither grounded on nor concerned with the horizon of its application, meaning its end. It is a practice of philosophy 'that is sought only as a science without having as its end anything more than the systematic unity of this knowledge, thus the logical perfection of cognition'.[95] The practitioner of scholastic philosophy thus runs the danger of becoming a *philodox*, who 'strives only for speculative knowledge, without looking to see how much the knowledge contributes to the final end of human reason'.[96] Consequently, scholastic philosophy is critiqued by Kant for caring about science for the sake of science and the logical perfection of this science alone, and ultimately being unconcerned with the purpose of this knowledge and its orientation within the world.

Philosophy in the world concept, on the other hand, is described as the 'science of the relation of all cognition to the essential ends of human reason (*teleologia rationis humanae*)'.[97] There are, then, explicitly *multiple* essential ends that philosophy in the world concept is concerned with.[98] However, these essential ends are themselves specified as means to yet another end, of which there is only one. This final end, Kant writes, is 'the entire vocation of human beings, and the philosophy of it is called moral philosophy'.[99] What differentiates scholastic philosophy from philosophy in the world concept in the *Critique of Pure Reason* is that the latter is concerned with more than the mere logical perfection of the system of cognition for its own sake. Rather, philosophy in the world concept aims to situate and generate the system in relation to the world as a system of essential ends – and then, in a second step, to relate this system of essential ends to the final end, that is the vocation of all human beings. Theoretical philosophy in the world concept, as a system of essential ends, is thus itself oriented towards yet another end, that is, moral philosophy as the 'science of the highest maxim for the use of our reason'.[100]

God, the world and the human in the *Opus Postumum*

Kant's architectonic, methodological reflections on the world and the different forms of knowledge resurface with renewed urgency within the context of the *Opus Postumum* and, more specifically, in relation to self-positing. The preceding discussion has already pointed out how reason 'makes' its ideas for the sake of the second act of self-positing. To this end, the *Opus Postumum* enacted an essential change to the threefold special metaphysics as Rational Psychology, the science of the soul, makes way for Anthropology, the science of the human. The preceding section has then shown that the human must, however, be understood as part of a complex cluster of systematic discussions, as anthropology forms only one part of what Kant calls knowledge of the world (*Weltkenntnis*), which is the necessary supplement to theoretical philosophy if theoretical philosophy aims to achieve more than a mere logical perfection of its system.

In the *Opus Postumum*, the reformulation of Rational Psychology into Anthropology takes place within the discussion of self-positing, itself concerned with the relation between transcendental and empirical self-consciousness and thus the process by which the subject 'makes' itself.[101] In a sense, however, the human already had something to do with self-making in the *Anthropology from a Pragmatic Point of View*. While anthropology from the physiological point of view 'concerns the investigation of what *nature* makes of the human being', pragmatic anthropology was always already described as 'the investigation of what *he* as a free-acting being makes of himself, or can and should make of himself'.[102] In the subsequent *Opus Postumum*, we meet this proposition again in the form of a question. Only now, it is transcendental philosophy rather than pragmatic anthropology, which is to answer the question 'What does man make out of himself?'[103]

What the human refers to here is, equally to God and the world, a mere idea. 'Each of these objects is absolutely one (*unicum*). If *God* is, he is only one. If there is a world in the metaphysical sense, then there is only one world; and if there is *man* he is the *ideal*, the archetype (*prototypon*), of a man adequate to duty.'[104] From this perspective, then, the human in question here 'is a mere idea of pure reason, the categorical imperative proves [*bewährt*] its reality, and it is therefore a noumenon'.[105] This means that the status of the existence of the idea of the human, as much as that of the world and God, can neither be affirmed nor denied. As an idea, it cannot be determinately answered from within the critical restrictions to knowledge set by Kant's critical project. At the same time and analogically to the world, the idea of the human is equally characterized as a task (*aufgegeben*), in the sense that it can never be given but is essentially subject to being brought about, '[. . .]the categorical imperative proves [*bewährt*] its reality'.[106]

This task is specified as a twofold undertaking in the following: 'I, the human being, am a world-being and an appearance in space and time.'[107] On the one hand, then, the human is here always already determined by its relationship to the world and as a being in the world.[108] At the same time to how 'man is, on the one hand, a world-being; on the other, however, man devoting himself to the law of duty: a noumenon [. . .]'.[109] Due to this double determination of the human, a trait that is without doubt familiar to readers of Kant's critical philosophy, Kant differentiates the idea of the human from that of the other two ideas, in a sense elevating and ennobling the idea of the human over and above the other two. Due to its twofold noumenal and phenomenal status, the human can take on the function of the copula between the ideas of God and the world, whereby mediating, combining and synthesizing the other two, otherwise heterogeneous ideas. This, the human can however only do according to the *Opus Postumum*, because the very distinction between God and the world is itself the result of the act of thinking. The human can only unify God and the world, because it is itself the power from which the distinction itself derives in the first place. 'The *medius terminus* (copula) in judgment is here the judging subject (the thinking world-being, man in the world). Subject, predicate, copula.'[110]

When taken together, the three ideas form a 'Gedankensystem',[111] and thus a system of thought rather than nature, 'similar to a science of forms, under which, if they were to be given, they would have to appear alone'.[112] Ultimately, this science of forms is nothing other than the system of transcendental philosophy that Kant had announced as early as the *Critique of Pure Reason*. While the *Critique of Pure Reason* self-reflexively announced itself as the 'science of the mere estimation of pure reason, of its sources and boundaries, as the propaedeutic to the system of pure reason',[113] it also put forward the plan for the entire system of pure reason, understood as 'the exhaustive analysis of all of human cognition'.[114] However, while announced as early as in the *Critique of Pure Reason*, it is only in the *Opus Postumum* that Kant finally ventured to lay open and unfold the long announced and sought-after system of transcendental philosophy.

Transcendental philosophy, as described in Fascicle I, is 'a science that presents (objectively) principles not about objects, but about the subject of cognition, about the domain and boundary of its knowledge'.[115] As such, it denotes more than the simple capacity of analytic reflection and self-limitation, and thus the world of analytic consciousness. Transcendental philosophy, Kant now proposes, is 'a faculty to represent oneself synthetically in one's thorough-going determination as given in a system of ideas a priori'.[116] Transcendental philosophy is thus explicitly articulated as the a priori presentation and constitution of oneself in the system of pure reason. Self-positing has thus, in the chronologically last Fascicle of the *Opus Postumum*, been reformulated as the self-constitution of reason to a system according to its ideas. '(Transcendental philosophy is the autonomy of ideas, insofar as they form, independently of everything empirical, an unconditional whole, and reason constitutes itself to the latter as a separate system.) God, world and the concept of the freedom of rational beings in the world.'[117] And, again, '[. . .] the thought through which we ourselves make these objects (through synthetic *a priori* cognitions from concepts) and, subjectively, are self-creators of the objects thought.'[118] A few pages later, Kant reiterates that 'Transcendental philosophy is the formal principle of systematically constituting oneself as object of cognition.'[119]

After the first, purely logical act of self-positing, in which the subject posits the forms of judgment, reason has entered the scene and constituted itself as a system of thought 'prior to its self-determination in space and time, yet for the purpose of the same.'[120] In Fascicle I, the problem of self-positing is thus put into relation with the question of the system of transcendental philosophy as a whole. Reason now posits itself by way of positing its three original and self-made ideas, which provide reason with its original images (*Urbild*). 'Transcendental philosophy precedes the assertion of things that are thought, as their archetype [*Urbild*], [the place] in which they must be set.'[121] These original images are necessary at this point and before any further act of positing, because, as Förster points out, 'without reason's ideal of a world, there would not even arise the notion of a sensible outer.'[122] Without reason's ideas, the very difference between the inside and the outside, and as a consequence, between the sensible and the intelligible, would not even arise. But how is one to understand this function of differentiation between the sensible and the intelligible that reason has acquired in the *Opus Postumum*, a function that in the *Critique of Pure Reason* was explicated as a problem of the concepts of reflection?

Reason, self-positing and transcendental reflection

The B-Edition of the *Critique of Pure Reason* argued that the determination of my consciousness in time essentially presupposes that I exist and occupy a position in space, a position that other things can therefore be external to. In the *Critique of Pure Reason*, however, how exactly such a position of externality is to be thought remained unresolved, since space, according to the transcendental idealist point of view, is the mere form of outer intuition rather than designating something 'outside' myself. Within the *Critique of Pure Reason* the concepts of the inner and outer are, properly speaking, concepts of reflection, the discussion of which is added as an appendix as the third and last part of the *Transcendental Analytic*, in a section entitled *On the amphiboly of concepts of reflection.*[123]

The mode of reflection (*Überlegung, reflexio*) introduced by Kant in the chapter in question is explicitly not concerned with objects themselves, as it rather concerns the 'state of mind [*Zustand des Gemüts*], in which we first prepare ourselves to find out the subjective conditions under which we can arrive at concepts'.[124] Furthermore, a difference between logical and transcendental reflection is introduced. Logical reflection compares concepts and establishes their identity and difference from one another. Transcendental reflection, on the other hand, has an altogether different function. Its primary responsibility is to sort any given representation and to distinguish whether it belongs to the faculty of understanding (as an object of thought) or the faculty of sensibility (as an object given in intuition):

> If it is not the logical forms but the content of concepts that is concerned, i.e., whether the things themselves are identical or different, in agreement or in opposition, etc., the things can have a twofold relation to our power of cognition, namely to sensibility and to understanding, yet it is this place [*Stelle*] **in which** they belong that concerns **how** they ought to belong to each other.[125]

Transcendental reflection thus establishes the relation between given representations by establishing the place (*Stelle*) and thus the faculty to which they belong. The important point, then, is that any judgment whatsoever essentially requires another judgment, that is, an act of transcendental reflection, in order to establish the 'transcendental place',[126] that is, the proper cognitive faculty to which each representation that is to be connected in judgment belongs.[127] The determination of such a transcendental place is explicated by Kant in reference to four conceptual couples of reflection: identity and difference, agreement and opposition, the inner and the outer, and matter and form.[128] Complementing transcendental place as 'the position that we assign to a concept either in sensibility or in pure understanding',[129] Kant further introduces the notion of a transcendental topic: 'In the same way, the estimation of this position that pertains to every concept in accordance with the difference in its use, and guidance for determining this place for all concepts in accordance with the rules, would be the transcendental topic.[130] The transcendental topic consists of these four conceptual couples of reflection, which '[precede] the concepts of things',[131] since they precede any judgment, be it synthetic or analytic. It is thus by way of transcendental reflection that the transcendental place of any representation must first be determined, from which, in a second step, judgments then becomes possible.

Differently to the *Critique of Pure Reason*, where the discussion of transcendental reflection was added on as an appendix to the *Transcendental Analytic*, the *Opus Postumum* now treats the problematic of transcendental place and topic as a problem of the transition from the first act of self-positing to the second. In the transition from logical to synthetic self-positing, reason puts forth a system of ideas called transcendental philosophy, which 'anticipates [*geht voraus*] the assertion of the things thought as their archetype [*Urbild*] into which they must be placed'.[132] The function of this original image (*Urbild*) put forth by reason is to establish the (transcendental) position (*die Stelle*) for any particular object of my thought, including of myself.

While transcendental reflection is not explicitly referred to in the sections in question, I would like to argue that the originary bifurcation, that is, the self-positing of reason as a system of ideas (God and the world), performs an analogical, primordial role of orienting reason according to the originary distinction between the sensible (phenomenal) and the intelligible (noumenal) world. For noumena, objects of pure understanding capable of being thought but never given, their transcendental place can only be in pure understanding, whereas phenomena as objects of sensibility can and must be given as appearances in sensibility.[133] It is only by way of establishing this initial distinction by way of two ideals, two proto-intuitive original images (*Urbilder*), that reason can orient itself and anticipate the place for each and every possible object of thought, including that of itself, in order to, then, determine itself further. And while in the *Critique of Pure Reason*, Kant does not specify how the power of judgment arrives at the concepts of reflection or the transcendental topic, the *Opus Postumum* now claims that the system of ideas is put forth by reason in the transition between the first and second act of self-positing.

Out of the first distinction of the totality of beings (*Das All der Wesen*) into God and the world,[134] forming an initial twofold transcendental topic, a cascade of further distinctions becomes possible, which in the *Critique of Pure Reason* were specified as the concepts of reflection, consisting of the inner and outer, matter and form, identity and difference, and agreement and opposition. As Förster proposed without, however, showing how this is to be done, I argue that reason essentially precedes the understanding in the *Opus Postumum*, in a similar way to how in the *Critique of the Power of Judgment*, reflecting judgment precedes determining judgment, because 'without its [reason's] projections thought could not orient itself in the world'.[135]

Technical-practical and moral-practical reason

The preceding discussion has argued that in the transition from the first to the second act of self-positing, reason makes itself into a system of thought according to its three ideas: the human, God, and the world. The resulting system of transcendental philosophy provides a transcendental topic, offering orientation to reason for the sake of its consecutive second, synthetic act of self-positing. Without the system of transcendental philosophy, thought could never find itself in the world and could also, in a further step, not constitute itself as an empirical object in the world.

The key to the second act of self-positing is, then, that the ideas of God and the world are 'ideas of moral-practical and technical-practical reason'.[136] Kant reformulates this point and articulates it in relation to transcendental philosophy and as an act of self-positing in another section which states the following: 'Transcendental philosophy is the act of consciousness whereby the subject becomes the originator of itself and, thereby, also of the whole object of technical-practical and moral-practical reason in one system – ordering all things in God, as in one system.'[137] The same point was first announced in Fascicle VII, before surfacing again and being treated in more detail in the very last Fascicle I:

> The subject determines itself (1) by technical-practical reason, (2) by moral-practical reason, and is itself an object of both. The world and God. The first is appearance in space and time. The second according to concepts of reason, that is, a principle of the categorical imperative.[138]
>
> God and the world. A system of transcendental philosophy of technical-theoretical and moral-practical reason.[139]
>
> God and the world. Each a unique whole according to quality. Singular and in connection. Technical/practical and moral/practical reason. Each a whole, and both together in combination (the totality of objects and of the subject).[140]

The beginning of this chapter saw Kant differentiate between the dynamic and mechanical moving forces of matter. The dynamic forces of matter, also called ether, were characterized by Kant as technical in Fascicle X of the *Opus Postumum*, which was furthermore said to make possible the whole of experience in the mechanical realm.[141] The original image (*Urbild*) of the world, as the successor of ether from Fascicle VII onwards, is now explicitly defined as the product of technical-practical reason. Consequently, this means that technical-practical reason does no longer only allow us to cognize the mere means necessary for a number of contingent ends and to bring forth objects in accordance with the laws of nature. In the *Opus Postumum*, Kant proposes that technical-practical reason also has a larger systematizing and orienting function, which together with moral-practical reason forms the absolute systematic whole of transcendental philosophy. Technical-practical reason is introduced to anticipate the whole of possible experience in the systematic whole called world. 'Space [*Weltraum*] is thought as a universal foundation [*Basis*] for the world of bodies [*Körperwelt*]',[142] meaning that objects of possible experience, here also already explicitly referred to as bodies (*Körper*), can a priori be given their place and relation to one another in the idea of the world as a system of ends not legislated by the categorical imperative, as they form, what is essentially, a cosmo-technical system.

While moral-practical reason brings forth and systematizes with the idea of God as 'a concept of reason of freedom, insofar as it contains a principle of the connection of the manifold that belongs to only one person',[143] technical-practical reason brings forth the idea of world as 'the whole of all objects of sensible representation'.[144] What differentiates these two systematic wholes is that reason, in each of these uses, 'makes' and systematizes according to a different set of ends (*Zwecke*). Moral-practical reason is oriented towards and legislates according to one final categorical imperative, while technical-practical reason concerns itself with all ends not legislated by the categorical imperative: 'The one relates to art and skill (for arbitrary ends), the other to wisdom, the final end.[145]

Importantly, then, technical-practical reason should not be read as subject to Kant's criticism of scholastic philosophy in the *Critique of Pure Reason* as done by Ottfried Höffe,[146] since scholastic philosophy is explicitly critiqued for its lack of concern for the foundation of knowledge and the horizon of application of knowledge in it aims for a mere logical perfection of the system of philosophy. As the preceding discussion has proposed, rather, it is precisely through technical-practical reason's contribution that theoretical reason's place in and relation to the whole called world can be established, a reading that is further supported by Gerhard Lehmann's claim in his *Habilitationsschrift*, according to which 'technical-practical reason [designates] precisely this world-reference of the theoretical reference to nature.[147] As such, and most importantly, technical-practical reason ensures nothing less than the world relation of theoretical reason and its cognitions, meaning that, without technical-practical reason's cosmo-technical contribution, theoretical cognition would be in danger of remaining unconcerned with the purpose and end of its cognitions, and thus its orientation, place and application within the world. As a consequence, then, both understanding and intuition, both the categorical apparatus and the forms of intuition, and, as such, the entire sphere of theoretical philosophy have been integrated into the technical-practical sphere. What remains to be shown, then, is how this second act of self-positing, at once technical-practical and moral-practical, will proceed according to the *Opus Postumum*.

16

Technical-practical self-positing

While Lehmann's observation that Kant did not explicitly write nearly half as much about technical-practical reason as about moral-practical reason is certainly correct,[1] the preceding section has laid out that, as a matter of fact, the entire discussion of the second act of self-positing in Fascicle VII must be understood as nothing but the process by which the subject determines itself as a sensible object among objects in the world, and thus as always already concerning technical-practical self-positing.[2] In Fascicle I, Kant offers the following reformulation of the second act of self-positing: 'The thinking subject creates for itself a world as the object of possible experience in space and time [. . .].'[3] World, the idea of technical-practical reason brought forth in the transition from the first to the second act of self-positing, is here said to be subject to further determination as the subject goes on to introduce, in the second act of self-positing, time and space, and thus the forms of intuition, into the world prepared by technical-practical reason. When it comes to Fascicle VII, we find the following sentence as early as in Beylage 2: 'It is not space as object which is intuited; space is, rather, the synthesis of the manifold in the representing subject itself. In this mode of representation, through which the subject constitutes itself [*breaks off*].'[4] The second step of self-positing consists in the synthesis of time and space as the manifold in the subject. Furthermore, however, the second part of this quotation shows how, for Kant, it is by way of synthesis that the subject posits itself as an object in space and time. Kant thus argues that the synthesis of time and space is in fact the process whereby the subject posits itself; time and space are forms of self-positing. 'Space and time in intuition are not things but *acts* of the power of representation positing itself, through which the subject makes itself into an object.'[5] A little further on, Kant repeats that 'The object of pure intuition, by means of which the subject posits itself, is infinite – namely space and time.'[6] Gerhard Lehman is thus right to point out that the determination of time and space as forms of self-positing is most certainly the most important development of the *Transcendental Aesthetic* within the *Opus Postumum*.[7]

By characterizing time and space as forms of self-positing, the *Opus Postumum* can say more about the source of the forms of intuition than the *Critique of Pure Reason* had previously conceded. The *Transcendental Aesthetics* famously laid out sensibility and understanding as two heterogeneous stems of cognition with an unknown common root,[8] propelling many a post-Kantian philosopher to embark on the search for this common root. With the discussion of self-positing in the *Opus Postumum*, Kant too now proposes that understanding and sensibility are no longer different stems of cognition with an unknown root. The two faculties derive from one common and underlying act, of which they are forms. Both the forms of the understanding and the forms of intuition are here, then, presented as forms of self-consciousness, and thus as complementary and necessary acts by way of which the subject posits itself.[9]

What is ultimately at stake in positing myself by way of positing time and space is that, according to Kant's transcendental restrictions to knowledge, the forms of intuition are our much-needed forms of object-receptivity, meaning that only by way of these forms can things be given to us in the form of appearances. The *Critique of Pure Reason* had defined appearances (*Erscheinungen*) as the 'undetermined [objects] of an empirical intuition',[10] which can be considered concerning both their form and matter. While the matter of appearance is said to 'correspond to sensation',[11] meaning that it is subject to being empirically given, its form is said to be what 'allows the manifold of appearance to be ordered in certain relations'.[12] In order to be able to carry out this function of ordering, the form of appearance can then itself not be given in sensation. Instead, pure space and time, as the form of appearances, are said to 'lie ready for the sensations a priori in the mind [*Gemüt*]'[13] to be activated, so to speak, by sensation.

When it comes to the *Opus Postumum*, the forms of intuition are still the subject's a priori forms of object-receptivity, and thus the only way in which the subject can be affected by objects, the only way by which things can be given to us. But differently to the *Critique of Pure Reason*, where the source of these a priori forms of intuition was left in enigmatic suspension, the *Opus Postumum* can now say more about where these forms come from. Already in Supplement 2, Kant writes that 'space and time are not things distinct from our representations as objects [*Gegenstände*] of our intuition, but they are intuitions themselves; yet not empirical (perceptions) of the sense-object, for in that case, they would be representations that come from another force that affects us, and as a cause, but they are products of our own power of representation'.[14] Time and space, Kant thus proposes, are products, and as such, they are subject to being made; they are *selbstgeschaffen* (self-made).[15] And the power that is said to 'make' these forms is here said to be our own power of representation (*Vorstellungskraft*).

Kant repeats that time and space are being made throughout the course of Fascicle VII, offering further insight into the powers that are involved in the spontaneous production of forms of object-receptivity. In Supplement 3, he writes, 'Space and time are products (but primitive products) of our own imagination, hence self-created intuitions, inasmuch as the subject affects itself and is thereby appearance, not thing [*Sache*] in itself'.[16] And in Supplement 4, he repeats his proposition that time and space are products of our own cognitive powers. But only Beylage 7 repeats that it is in fact the power of imagination that 'makes' the forms of intuition.

'It's representation [space and time] is an act of the subject itself and a product of the power of imagination for the sense of the subject.[17] In other places, he returns to charging the power of representation as a whole with making the a priori forms of intuition, as in Supplement 4, for instance, when he writes

> Space and time are not discursive representations through concepts, but intuitive representations. They are not objects [*Gegenstände*] of the intuition of the manifold, but intuition itself, and specifically, not of the universal, but of the singular (there is one space and one time). It is not a perceptible object (apprehensibile), but a product of the faculty of representation as self-activity (Spontaneitas not Receptivitas) to represent the aspectabile as cogitabile.[18]

And a few lines further, he writes, 'Space and time are not perceivable (*apprehensibile*) objects that exist outside of my representation, but they are themselves creations [*Geschöpfe*] of my faculty of representation.[19]

There is then, an a priori making of the forms of time and space. What 'making' these a priori forms refers to is an a priori synthesis (*Zusammensetzung*). The *Critique of Pure Reason* had claimed that the synthesis of apprehension 'must be exercised *a priori*, i.e., in regard to representations that are not empirical'.[20] And indeed, by saying that the a priori forms of time and space are posited and brought fourth (*selbstgemacht*) by the power of imagination, Kant says as much as that the pure forms of time and space are subject to synthesis. In calling this production an originary act, Kant decisively refers to the productive, rather than the reproductive power of imagination, and thus pure synthesis. Pure synthesis apprehends and comprehends the pure manifold of space-time itself rather than any given empirical manifold, meaning that pure synthesis deals with our forms of affectability themselves.

In the *Opus Postumum*, however, rather than talking about apprehension and comprehension and thus what together makes up synthesis according to the *Critique of Pure Reason*, Kant specifies the work of the productive power of imagination in the act of positing space and time in a wholly other manner. If one searches the *Opus Postumum* for the specific 'mode of production' of the forms of intuition by the power of imagination, one finds a number of passages like the following:

> The understanding does not start from the object, but from its own subject, in order to construct the sensible intuition, according to its form; that is, to present the manifold of sensible intuition synthetically *a priori*, in the unity of the manifold, according to a principle – which is a mathematical operation of the understanding, and an act of transcendental philosophy [. . .]. The unconditioned unity of the manifold in intuition is not *given* to the subject by another object, but is *thought* through itself. Space and time are not anticipations of perception, as concepts of the understanding, but forms of the objects in appearance.[21]

As an a priori construction, time and space are thus explicated in distinction to the anticipations of perception, which according to the *Critique of Pure Reason* determined that the real of perception can be anticipated according to its degree of intensity alone.[22] Kant's assertion that the power of imagination synthetically constructs the pure forms of intuition is made not only in Fascicles X and XI but also appears in the later Fascicle VII, where Kant writes that 'what is formal in this intuition is One and All, coordinated; [it] is the representation of space and time, which represents an infinity (unlimited magnitude), not analytically through concepts, but synthetically through the construction of concepts'.[23] As forms of intuition, Kant writes, their representation

> is pure not empirical intuition, which does not constitute itself like in logic from the universal to the particular analytically, but rather synthetically constitutes itself a priori from the totality of the sum total of the manifold to the singular and contains only the formal aspect of the manifold of the intuition of the subject in its totality, that is, the unconditioned whole of intuition.[24]

This means that we construct the pure forms of intuition purposively into the unconditioned whole of intuition as an infinite magnitude.

Furthermore, that which is properly said to carry out this act of construction by means of the power of imagination is technical-practical reason: 'There is, however, apart from sensible representation, yet another faculty of knowledge, which contains not merely receptivity but also spontaneity (as highest faculty of knowledge): namely, understanding, judgment and reason. The latter can be either technical, intuition-constructing reason or moral-practical reason, both combining *a priori* the manifold of representations to knowledge under a principle'.[25] This act of a priori construction is thus explicated as a spontaneous act of technical-practical reason, which is here said to enlist the power of imagination to construct itself into an object of intuition within the whole of all appearances. Lehmann is thus right to argue that the entire formal realm of appearances is subsumed under the teleological, technical sphere as a product of technical-practical reason.[26]

The problem of construction

Kant's determination of the pure forms of intuition as the technical-practical product of intuitive construction brings up a number of problems that threaten to transgress the critical limits set during the critical period by Kant himself. The *Critique of Pure Reason* was organized around a founding methodological distinction between philosophy and mathematics. Mathematics constructs its concepts in intuition a priori, meaning that it can provide intuitions to its concepts through mere a priori presentation (*hypotyposis*, *Darstellung*). Kant's 1790 text *On a discovery whereby any new critique of pure reason is to be made superfluous by an older one*[27] defined construction as the 'exhibition of a concept through the (spontaneous) production of a corresponding intuition'.[28] This general sense of construction was then further distinguished into pure and schematic construction and empirical construction, which was also called technical. While pure construction was specified as occurring 'through mere imagination in accordance with an a priori concept', technical construction, which was further differentiated into geometrical and mechanical construction, was said to be 'carried out on some kind of material'.[29] In the *Critique of Pure Reason* Kant made allowances for mathematics to demonstrate its concepts through schematic construction since mathematics only deals with the concept of magnitude to begin with, which contains 'just that which the definition would think through it'.[30] As such, the concepts of mathematics can be constructed '*in concreto* and yet a priori'.[31] This means that in mathematical construction, the validity of a concept can be determinately cognized a priori through a priori intuition alone, and thus without taking recourse to empirical data. If the concept (of construction) were to have any relation to a possible content of intuition whatsoever, meaning if the concept 'contain[ed] an existence and correspond[ed] to sensation',[32] such a priori presentation would be illegitimate, however, since such content can be given only in empirical actuality and thus a posteriori.

Consequently, the *Critique of Pure Reason* upheld a difference between mathematics and philosophy since, distinct from mathematical concepts, philosophical concepts are subject to schematization rather than construction, both of which are types of *hypotyposis* (presentation). Importantly, the objective validity of concepts can also be given a priori through schematization, but only insofar as schematization offers 'a general but sufficient characterization of the conditions under which objects in harmony with those concepts can be given'.[33] Philosophical concepts gain their validity in relation to the conditions of possible experience, which extend beyond concepts of magnitude, and due to this, Kant vehemently argued against the application of the geometrical-mathematical method in philosophy.[34]

When it comes to the *Opus Postumum*, Kant designates the a priori construction of intuition as the technical-practical second act of self-positing, whereby transgressing his own prior methodological distinction. What was hitherto relegated to technical construction alone has been elevated into the transcendental, whereby invalidating the earlier distinction between schematic and technical construction. The second step of technical-practical self-positing begins with nothing less than the intuitive construction (a priori presentation) of the concept of the subject in intuition. Since these forms of intuition are, however, spontaneously constructed by the power of imagination, they can only present the infinite magnitude of intuition *in concreto* and a priori, without any relation to the reality of that which appears in space and time, meaning myself as an object of existence.

At the same time, however, Kant must have deemed it possible to avoid his own critique of Fichte's purely logical system of thought, which he denounced as incapable of ever relating to any real objects. This means that something more is needed. In *Kant's Final Synthesis*, Eckhart Förster methodologically follows Kant's account of self-positing and extends it to a total of five, rather than merely two acts. Förster's interpretation and amplification of self-positing, which demands much freedom of interpretation with regard to the *Opus Postumum*, proposes that a supposed third act of self-positing is needed to determine my own existence as a corporeal organic being in space and time by realizing space as an object of the senses.[35] He thus argues that there is a third act which proceeds by way of 'the assumption of a universally distributed ether', and thus the hypostatization of space, and that it is this third act which will provide the forms of intuition with it's a priori content.[36]

Stephen Howard critiques Förster's five-step interpretation by arguing that self-positing is the mere subjective part of the transition project, which must be read in its essential relation to the objective pole, which he finds in the ether proofs.[37] Critiquing Förster's account, Howard then goes on to argue that what is needed is a bridge between the objective and subjective poles, since without such a bridge, neither of the modes of self-positing by themselves would ultimately suffice. Howard finds this bridge in Fascicles X and XI. I agree with Howard's critique of Förster, in the sense that the intricate series of steps in Förster's account treats self-positing in quasi-isolation and thus does not account for the dynamic nature of Kant's thought and how problems are addressed by Kant across different Fascicles and differing concepts. However, differently to Howard, who reads Fascicle VII in relation to the earlier Fascicles X and XI and thus in relation to Kant's renewed engagement with physics, I have in the preceding discussion moved from Fascicle VII to the very last Fascicle I. While Howard identified the objective pole as ether, I have argued that Kant's engagement with ether will transform into his engagement with the idea of world in Fascicle I towards the end of the manuscript.[38]

Similar to Howard, then, who argues for the necessity of a transition within the transition project of the *Opus Postumum* as a whole, I have traced in the preceding pages that and how there is a transition in the sense of a progression between the two acts of self-positing. And it is this very transition from the first to the second act of self-positing which ensures reason's relation to the world and, furthermore, position in a world of objects. How, then, is it possible for Kant to avoid his own critique of Fichte and move from mathematics to philosophy, and thus from the mere a priori construction of time and space as infinite magnitudes to the consciousness of myself as a being in the world by way of this transition, and what role does technical-practical reason play here?

Self-affection and appearances of appearances

The *Critique of Pure Reason* had famously introduced the idea that in the a priori synthesis of time and space, the subject comes to affect itself. This is due to the fact that the forms of intuition are themselves nothing but the subject's forms of affectability (or object-receptivity). And while space is the way in which things can be given to me, time is the way in which I can be given to myself. As pure forms of intuition, however, time contains only the relations of succession, simultaneity and persistence in time, while space contains the relations of extension, motion and moving forces in space.[39] For Kant, this means that 'through mere relations no thing in itself is cognized',[40] meaning that the a priori form of space only concerns the place (*Ort*) of something or its change of place but cannot assert anything concerning what is given and real (*gegenwärtig*) in that place. The same applies to the a priori form of time, which concerns the temporal relations in which a representation can be set. Consequently, the pure form of time 'does not represent anything except insofar as something is posited in the mind [*Gemüt*]'.[41] At the same time, however, and in this very activity of positing the relations of time in the mind, 'the mind [*Gemüt*] is affected by its own activity, [. . .] it is affected through itself'.[42] According to the *Critique of Pure Reason*, then, the subject comes to affect itself through the a priori synthesis of time and space, through which it can, in a second step, know something more. And that something is myself, not how I am in itself, but how I appear to myself.

Following the basic outline of self-affection put forth by the *Critique of Pure Reason*, the *Opus Postumum* proposes that 'our sensible intuition is, initially, not perception (empirical representation with consciousness), for a principle of positing oneself and of becoming conscious of this position precedes it; and the forms of this positing of the manifold, as thoroughly combined, are the pure intuitions, which are called space and time'.[43] In the act of construction of the pure forms of intuition, the subject is thus said to become conscious of itself: '[. . .] the composing subject appears to itself in the composition according to principles'.[44] The subject is thus also here said to be at once the cause of affection and that which is affected, at once the active cause of construction of the forms of self-positing and at the same time that which appears to itself in this construction. But, differently to Kant's claim in the *Critique of Pure Reason*, where the subject was said to appear to itself just 'like other phenomena',[45] the subject in the *Opus Postumum* no longer appears to itself just like any other phenomenon. In fact, the subject is now said to appear to itself unlike anything else. Introducing an inner distinction into appearances, Kant now claims that the subject appears to itself as an appearance of an appearance.

With the *Opus Postumum*, Kant's thought on appearances (*Erscheinungen*) has thus undergone a fundamental change. Appearances have now become twofold, in the sense of being internally differentiated into first and second-order appearances, which Kant also calls direct and indirect appearances, with the latter also called appearances of appearances (*Erscheinungen von der Erscheinung*). Appearances of the first order, or direct appearances, represent objects in relation to how the subject is affected by them and thus as the way in which (undetermined) objects of empirical intuition are given to us. Appearances of the first order are claimed by Kant to refer to the objective, physiological mode of representation.[46] The subject that appears to itself, however, is said to be an altogether different kind of appearance. It is the 'appearance of the self-affecting subject (hence indirect)',[47] and thus the way that the subject appears to itself and becomes conscious of its own activity of self-positing, of combining and unifying its inner perceptions, the effects of its own moving forces into a whole, a system whereby it then, in a second step, appears to itself as this system, as a special object, unlike any other appearance, that is, an appearance of an appearance.[48]

In many ways, the discussion of self-affection is the precursor to self-positing. Self-affection and its corresponding notion of the dual order of appearances are first discussed in Fascicles X and XI, written between August 1799 and April 1800, and thus in the period immediately preceding Kant's engagement with self-positing. In scholarly literature on self-affection, particular attention has been paid to the question of whether, in his final work, Kant conceded there to be a double-affection at play, with the empirical object affecting the empirical subject and the thing-in-itself acting on the transcendental subject. Both Adickes, Kemp Smith and Vaihinger read Fascicles X and XI as Kant's late concession to double-affection.[49] Lehmann, on the other hand, reads self-affection in the *Opus Postumum* as a final testament to Kant's abolition of the thing in itself, interpreting self-affection and affection as two sides of the same coin.[50] My focus in the remainder of this section, however, is limited in scope. I will not intervene in the systematic discussions on self-affection in the *Opus Postumum* put forward by readers like Adickes, Lehmann and Förster, as it is ultimately not the question of this project. Rather, the remainder of this chapter will inquire into the relation between appearances of appearances and technical-practical reason, that is, the question how does, what first appeared under the name of second-order appearances in Fascicles X and XI, relate to technical-practical reason and its corresponding original image (*Urbild*) of the world.

The preceding discussion has shown how technical-practical reason engages in the act of constructing the forms of intuition. In the a priori synthetic presentation of the concept of magnitude in intuition, the subject is said to affect itself and, through this affection, be given to itself as an appearance of an appearance. Appearing to oneself as a second-order appearance means that I appear to myself in the act of spontaneously constituting myself into an object by means of the construction of pure intuition. While a first-order appearance is the 'undetermined [object] of an empirical intuition',[51] and thus the representation of an object of sense ordered and placed in intuition but not yet thought according to the unity of the categories, the second-order appearance only concerns the formal relations of an object to the subject, rather than any given empirical intuition of an object, and thus includes the subject of appearances itself as part of that which appears. To appear as 'part' of the world of appearance, however, already means that the subject is now located in and occupies a position within the world of appearances, albeit at this point only formally. And only because the subject is located in and occupies a position within the world of appearances can the subject then have appearances of the first order.[52]

What thus appears under the name of a second-order appearance is the technical-practical subject, which posits itself by positing the pure forms of time and space in order to be able to be given objects in a second step. This means, however, that the transcendental forms of intuition and understanding, and thus that which the subject contributes (posits, makes, brings forth) for the cognition of given empirical things, are now able to appear to the subject under the name of an appearance of an appearance.[53] And it is in this sense that Kant, in a second step, writes that 'appearance of an appearance is that through which the subjective is made objective'.[54]

An appearance of an appearance is thus ultimately the reflective self-consciousness of the technical-practical subject in the process of positing itself in the act of synthesis of space and time for the purpose of orienting itself within and cognizing the world. Thus, while we cannot learn anything about any given thing by way of second-order appearances, what we do learn about is our relation to any giveable thing and our place in the world as a whole. Technical-practical self-positing determines the world as *mundus*,[55] the world 'in space and time [. . .] together with the subject (the human), which itself is a part of the world'.[56] And the human that appears to itself as part of this spatio-temporally determined world is what Lehmann called the technical-practical subject.[57]

17

Conclusion

Thinking back to the beginning of this chapter, it becomes clear that technical-practical reason was subject to a radical transformation between its initial articulation in Kant's moral philosophy and the *Opus Postumum*. Technical-practical reason first appeared in relation to technical-practical imperatives, and thus rules of skill and counsels of prudence,[1] as it concerned purposive actions not explicitly legislated by the categorical imperative. While already insisting that, properly speaking, these modes of purposiveness are to be dealt with internally to theoretical philosophy, a claim repeated by the *Critique of the Power of Judgment*,[2] the relation between technical-practical reason and theoretical knowledge was at once discussed as a problem of mere application with the idea of a technics of nature introduced as a further condition of possibility for science.

As the preceding chapter has shown, technical-practical reason in the *Opus Postumum* is responsible for legislating one part of the cosmological universe of all things constituted by reason in its unity. Bifurcating into moral-practical reason on the one side and technical-practical reason on the other, reason in this twofold use was shown to be productive of ideas that function as original images (*Urbilder*) and provide the transcendental place of the sensible and the intelligible. Technical-practical reason brings forth the idea of world, which provides the stage and ground of the whole of all objects of sensible representation that can be given to us, while moral-practical reason brings forth the idea of God, which concerns the concepts of freedom, duty and right.

Similarly to transcendental reflection, which searched for the belonging of any given representation to either pure reason or sensible intuition by means of the concepts of reflection, reason here introduces a fundamental distinction into the totality of beings (*Das All der Wesen*).[3] After the bifurcation of reason into the two wholes of God and the world, self-positing is subject to a twofold treatment from here on by moral-practical reason on the one hand and technical-practical reason on the other, the second of which concerns the self-constitution of the human within the world.

I have argued, then, that for the *Opus Postumum*, the subject of cognition is essentially technical-practical, as it 'creates the elements of knowledge of the world himself, a priori, from which he, as, at the same time, an inhabitant of the world, constructs a world-vision [*Weltbeschauung*] in the idea'.[4] The technical-practical subject thus first makes, posits and synthesizes an archetype of this world 'in his own self'.[5] In the context of this effort of knowing the world by way of making it first in one's own self, Kant calls this technical-practical actor *cosmotheoros*, a rare Greek neologism that Kant most likely adopted from Christian Huygens's 1698 book of the same title. The *cosmotheoros* a priori makes the elements for the theoretical cognition of the world in its own self.

Consequently, while the *Critique of Pure Reason* had famously held the position that philosophy 'is the science of the relation of all cognition to the essential ends of human reason (*teleologia rationis humane*), and the philosopher is not an artist of reason but the legislator of human reason',[6] the *cosmotheoros* of the *Opus Postumum* is now very much an 'artist of reason', meaning that the *cosmotheoros* could also carry the name of *cosmotechnikos*, an artist or technician of the world. Differently to the total injunction against theoretical reason's talent of building, expansion and objectivation (*Äußerung*) imposed in the *Critique of Pure Reason*, the *Opus Postumum* thus critically delineates the conditions of possibility under which technical-practical reason can both produce its own elements of knowledge and methodologically employ them. Having decisively elevated technical-practical reason from a problem of empirical philosophy without a need for a critique, to a transcendental faculty in its own right, the *Opus Postumum* thus indeed, as first proposed by Gerhard Lehmann, contains the outline to what would have been Kant's third critique of reason, that is, a critique of technical-practical reason.[7] As such, it is my contention that this trajectory from Kant's critical restriction of instrumental reason at the outset of the *Critique of Pure Reason*, followed by a slow, continued development of a critique of technical-practical reason in its own right, puts forth a decisive image of Kant as a thinker of technics.

While, on the one hand, this image of Kant as a thinker of technics challenges Kantian and post-Kantian philosophy in its omission of the meaning and role of technics in Kant's Transcendental Philosophy *tout court*, it also troubles the account of Kant put forward in the field of philosophy of technics, exemplified by the discussions of Heidegger, Simondon, and Stiegler. As Chapter 1 has shown, Heidegger, Simondon, and Stiegler essentially read the history of philosophy as the history of philosophy's technical oblivion and transcendental philosophy as philosophy's final obstacle to thinking technics, despite finding the seeds to their own technics-thinking in their engagement with and reading of Kant.

In opposition to Heidegger, Simondon, and Stiegler, I have argued that Kant explicitly read all philosophy prior to the *Critique of Pure Reason* as the history of instrumental reason, while he at the same time outlined the conditions of possibility for technical-practical reason, putting forward a transcendental concept of technics. Against Heidegger, Simondon, and Stiegler who diagnosed technics as Kant's glaring and systematic omission and thus as the one thing that Kant was structurally incapable of seeing, I have argued that the *Opus Postumum* explicitly carves out a space for technics in its transcendental form in the realm of appearances. Thus, what was critiqued by Heidegger, Simondon, and Stiegler as Kant's symptomatic omission is here explicitly investigated and laid open under the name of appearances of appearance. How so?

First introduced as providing the transition (*Übergang*) between metaphysics and physics in the *Opus Postumum*, the idea of world is heir to what had hitherto been investigated under the name of ether by Kant. Similarly to ether, the idea of world is needed to provide an idea in which and through which a particular object of experience can be given and placed. Determining the original image of world further by means of positing time and space, the technical-practical subject then appears to itself in this act of self-positing through self-affection. While impossible within the first order of appearances, the conditions under which things can be given to us as things – the product of technical-practical self-positing – can now appear to us precisely under the name of second-order appearances.

My reading of the *Opus Postumum* thus challenged the narrow account of technics put forward by Heidegger, Simondon, and Stiegler, which essentially consists of a pre-industrial and thus artisanal *techne*, and an industrial, modern instantiation. Kant's account of technical-practical reason problematizes and challenges these narrow, historically simple and universal notions of technics put forth by Heidegger, Simondon, and Stiegler as it falls outside the simple twofold schema. Kant engaged two parallel genealogies of technics at once, the medieval scholastic notion of *Technologie* and the Aristotelian notion of *techne*. That he further shortened the term *Technologie* into the modernized German term *Technik* further shows that Kant was aware of Beckmann's invention of the modern science of *Technologie* as Kant aimed to distance his (transcendental) discourse on technics from the applied science of technology. Since essentially, what is at stake in the question on the relation between Kant and technics is, however, the much larger, encompassing question about the relation between philosophy and technics, I have shown that there furthermore exist philosophical resources for thinking technics prior to Heidegger, Simondon, and Stiegler's interventions, despite their failure to recognize them.

Finally, then, Kant's reformulation and critique of technical-practical reason in the *Opus Postumum* shows most clearly that he discussed technics as essentially part and parcel of a larger cosmological problematic. In surprising fashion, Kant's technics-thinking must essentially be understood as the hitherto eclipsed quasi-predecessor of contemporary cosmo-technical debates that have since emerged out of Simondonian thought and are the subject of work from Isabelle Stengers, through Bruno Latour and onto Yuk Hui. Yuk Hui has argued in *The Question Concerning Technology in China* that at the root of the majority of twentieth century discourses on technics lies an anthropologically universal concept of technics, which means that technics is read as a universal principle that transcends geographical and historical specificities.[8] His notion of cosmo-technics aims to critique the widespread hypothesis of technology as such an anthropological universal. Instead, he aims to show how concepts like *techne*, *physis*, or *metaphysics* in fact receive their meaning and are translated in relation to their position within a larger cosmological system.[9] Consequently, he argues that technology must not be understood as an anthropological universal, as it rather gains its definition and meaning in relation to a cosmo-technically specific past and present.

Kant's essential and fundamental contribution to cosmology in the *Critique of Pure Reason* was to reconfigure the world as given into the world as a task for reason and thus as *aufgegeben*. I have shown that in the *Opus Postumum*, the faculty at once charged with this task and charging itself with the task is technical-practical reason. What is specific and ultimately promising about Kant's account here, then, is that Kant argues that, and conceptualizes how any system with its inherent characteristics of universality and necessity only becomes possible within a world built by technical-practical reason. For Kant, science, technology, or any other system of necessity always already sets off in a world built by technical-practical reason. According to Kant, the world of science is thus a technical-practical projection, in as much as the world of technics in the empirical sense is equally situated in this larger cosmo-technical whole, which is the first and most important responsibility of technical-practical reason.

Furthermore, however, the world built by technical-practical reason is also the (only) world in which the human can appear to itself as a technical-practical subject, and thus as at once free and restrained by its relation to the world. Finally, then there can indeed be no access to the world outside or before technical-practical reason according to the Kantian schema, since the world is not a problem of aesthetics nor of any other pre- or post-rational modes of knowledge and engagement. Instead, the problem of the world is the fundamental task given to itself by technical-practical reason, as cosmology in the *Opus Postumum* essentially becomes a problem of cosmo-technics.

Epilogue

Kant, forgetting, and technics, again

The context of my preceding inquiry into Kant and technics has been provided by a specific discourse. I have laid out how Heidegger, Simondon, and Stiegler all argued that since its beginnings in ancient Greece, philosophy in its canonical form had constituted itself by way of distorting, repressing, and ultimately forgetting technics. Built on the exclusion of technics, philosophy was consequently diagnosed by Heidegger, Simondon, and Stiegler as being structurally incapable of seeing, understanding, or thinking technics. At the same time, technics was not only characterized as philosophy's ultimate obstacle. According to Heidegger, Simondon, and Stiegler technics also posed philosophy's ultimate possibility: if philosophy is to play a role in what has without doubt become a technological world, then it must reconstitute itself anew, only this time in its essential relation to technics. To think technics, according to Heidegger, Simondon, and Stiegler, thus concerns nothing less than the future of philosophy as a whole.

At the centre of philosophy's technical exclusion, oblivion, and repression, Heidegger, Simondon, and Stiegler positioned Kant as modern philosophy's most prominent techno-problematic representative. The preceding chapters have shown that Kant's role in this discourse was however paradoxical. On the one hand, the philosophical task of thinking technics for Heidegger, Simondon, and Stiegler essentially meant overcoming the transcendental limits to knowledge so forcefully articulated by Kant, since transcendental philosophy was diagnosed as philosophy's limit of thinking technics. On the other hand, I have shown that Heidegger, Simondon, and Stiegler's own technical thought relied, each in its own way, on a problematic reading of Kant in which Kant implicitly articulated fundamental aspects of the modern concepts of technics, without, however, self-reflexively understanding the nature of his own insight. The explicit aim of the preceding chapters was not only to show that Kant explicitly thought and wrote about technics but to furthermore trace and illuminate the transformation that his technics-thinking underwent from the 1781 *Critique of Pure Reason* until the

Opus Postumum, the posthumously published manuscript that Kant worked on until shortly before his death in 1804.

However, as I made my way through the pages of the *Opus Postumum*, I realized that I had arrived at yet another scene in which philosophy, technics, and forgetting have come together in a problematic disputation. The *Opus Postumum*, in which Kant also most explicitly and frequently writes about technical-practical reason, has, ever since Kant's death, slowly become the scene of philosophy's long and manifold attempts of immunizing itself against the threat posed by forgetting, which is here Kant's forgetting. We are told by numerous sources that at the end of his life, Kant suffered from substantial memory decline; he had become senile. When one makes one's way through the manuscript of the *Opus Postumum*, is one thus witness to the evolution of Transcendental Philosophy in Kant's daily practice of philosophizing, or rather to its atrophy? Is the *Opus Postumum* the product of Kant's merciless but rational work of rearticulating his earlier critical positions first outlined in the 1780s, or are the changes introduced rather the irrational product of Kant's senile forgetfulness?

The question of Kant, technics, and forgetting thus presents itself once again, albeit in a different form. Now, it concerns the question and problem of Kant's forgetting during his last years and thus during the writing period of the *Opus Postumum*, including his final engagement with technics. The remaining pages will attempt to shed light on this new conjunction.

Styx and Lethe

As pointed out by Stephen Howard, most post-war scholarship interprets Kant's commencement on the manuscript of the *Opus Postumum* in relation to two letters in which Kant reports of his realization that his philosophical system was endangered by a gap (*Lücke*).[1] First mentioned in a letter to Christian Garve in 1798,[2] the existence or non-existence of this gap, however, remains disputed. Some readers of the manuscript embarked on identifying the threatening gap, while others rejected the very proposition of there being such a gap on the argument of the sufficient systematicity of the critical system as outlined in the 1780s.[3] While the question of the gap, including its relation to the transition provided by the manuscript (most often Kant referred to the manuscript as *Übergang von den metaphysischen Anfangsgründen der Naturwissenschaft zur Physik* and thus *Transition from the Metaphysical Foundations of Natural Science to Physics*), is the subject of both historical and ongoing debate,[4] what is nevertheless clear is that as Kant's work on the manuscript proceeded, the way in which the transition was to be carried out itself became subject to transformation until his work on the manuscript eventually ceased sometime around the year 1803 before Kant's eventual passing on 12 February 1804. As such, the end of the manuscript is not its own. It is determined, perhaps overdetermined, by the end of Kant's life. And thus, what was meant, at last, to complete Kant's system was itself never completed.[5]

Or perhaps, one might say, that it was in fact completed. The transition is likened by Kant to the river Styx at various points throughout the manuscript.[6] According to ancient Greek mythology passed down to us in Hesiod's *Theogony*, Styx was the daughter of Ocean and Tethys, who, in the war of the titans, was the first to rush to Zeus' aid.[7] Consequently, Zeus named Styx the goddess of oath from when on every oath, even that of the gods, must be sworn by her.[8] The oath sworn by the river Styx has, since the most ancients, been charged with the task of bridging the gap between thinking and being, that is, the name and the thing named, and is invoked by Kant for the same purpose in his transition project.[9] The crossing of the river Styx is however also the final journey that every human must eventually embark on, pointing to the radical finitude of the human, which is here Kant's finitude. With the turning of each page of the manuscript, we accompany Kant in his daily practice of philosophizing until work on the manuscript eventually ceased. Rather than considering the *Opus Postumum* an incomplete work, it could thus perhaps be more fitting to say that the transition, that is the journey over the river Styx, has been completed.[10]

Styx was only one of a total of five rivers of the ancient Greek myth of the underworld.[11] Next to Acheron, Styx, Phlegeton and Kokytos, it was the role of the river Lethe to bestow forgetting upon the souls of the deceased.[12] According to Kant's three first biographers, Ludwig E. Borowski, Reinhold B. Jachmann and E. A. Ch. Wasianski, often referred to as the Nicolovius-biographers,[13] the redeeming forgetting administered by Lethe had however tragically set in prior to death in Kant's case. According to Jachmann, a memory decline had started as early as 1795, prompting Kant to increasingly use notes to aid his memory over the remaining years of his life.[14] According to Wasianski's recollections, Kant had been forgetful throughout the course of his entire life. Retelling a story from Kant's childhood in which he had once forgotten his schoolbooks, Wasianski claims that Kant had in fact himself admitted at various occasions that his memory had never been one of his strengths.[15] And what had always been bad, only got worse over Kant's last years. Differently to Jachmann, the onset of Kant's senile, rather than 'usual' forgetfulness is, however, dated by Wasianski to the later year 1802. Since also Borowski, the third of the Nicolovius-biographers, reports that Kant had in fact always been forgetful,[16] it thus indeed appears that Kant's memory might never have been one of his strengths. A late, further decline of Kant's power of memory is nevertheless diagnosed by all three biographers, while the onset of the decline is dated differently in each of their accounts.

The reception history of the *Opus Postumum* has been tightly interwoven with and influenced by Kant's reported senility. Wasianski, who also functioned as Kant's executor following his death, professed in his biography that while at points, Kant referred to the manuscript as nearly finished, he also wanted that same manuscript to be burnt after his death. Wasianski took this to mean that Kant's mental powers had, due to his increasing senility, so diminished over the last years of his life that he had become incapable of judging the merit and importance of his own work.[17] Given Kant's own lack of judgment, Wasianski handed the manuscript to Johann Friedrich Schultz for examination, who promptly advised Wasianski against the publication of the manuscript. Not even the introduction was finished, Schultz claimed, making a posthumous edit or completion impossible.[18] Having gotten external confirmation from Schulz, Wasianski decided against publishing the manuscript left behind by Kant. However, as Giovanni Pietro Basile points out, Wasianski did more than merely decide against the edition and publication of Kant's last manuscript. Most likely in an attempt to protect Kant and his philosophical legacy, Wasianski made the manuscript disappear completely.[19]

Following this first fateful assessment by the three Nicolovius-biographers, the prominent nineteenth-century historian of philosophy Kuno Fischer published his dismissive assessment of Kant's last manuscript in his 1860 *Geschichte der neuern Philosophie*. Without ever inspecting the manuscript itself, Fischer polemically argued that the manuscript should be ignored as a whole, since it is testament to Kant's mental 'deteriorating condition'[20] at the time of writing. A few years later, he repeated his first diagnosis, referring to the late Kant as a 'decaying spirit'.[21] Fischer was, however, not alone in interpreting the writing of the late Kant as the product of senility. Maja Schepelmann points to a whole culture of Kant interpretation, in which senility, first mentioned by the three Nicolovius-biographers, is employed as an explanatory framework for difficult passages in Kant's texts.[22] Hans Vaihinger already took recourse to a '*Blattversetzungsthese*'[23] (leaf displacement theory) in relation to Kant's 1784 *Prolegomena*, the difficulties of which, he argued, can only be understood as consequences of Kant's failure to notice that the order of his text had been corrupted in the editing or printing process. According to Schopenhauer, Kant's *Doctrine of Right* and thus the second part of his 1797 *Metaphysics of Morals* is clearly a product of Kant's senility.[24] In the privacy of a letter to Karl Rosenkranz in 1837, which was however printed one year later in the introduction to the 1838 Rosenkranz edition of Kant's *Critique of Pure Reason*, Schopenhauer went further in arguing that the changes undertaken by Kant in the 1787 B-Edition of the *Critique of Pure Reason* are the product of Kant's *Altersschwäche* (senility).[25] In contemporary Kant scholarship, neither the 1784 *Prolegomena* nor the 1787 B-Edition of the *Critique of Pure Reason* are still read as products of, or compromised by, Kant's senility. But Schepelmann points out that a similar logic persists in reference to Kant's late philosophy. Similarly to Vaihinger, Bernd Ludwig has proposed that the diminished legibility and comprehension afforded by Kant's 1797 *Doctrine of Right* might be the result of editing mistakes, which the ageing Kant no longer seemed to notice.[26] When it comes to the 1798 publication of Kant's anthropology lectures under the name *Anthropology from a Pragmatic Point of View*, no sign of the conceptual force of Kant's earlier philosophy remains, according to Reinhard Brandt's evaluation.[27] And in his 2003 Kant biography, Manfred Kühn concludes that Kant simply did not possess the energy and power of his earlier self in his late writings. A decline in Kant's mental powers was first noted by Kühn with the 1798 *Conflict of the Faculties*.[28]

While there is clearly a larger 'culture' of explaining difficult passages in the text through Kant's supposedly diminished mental powers, this argument can nowhere be observed more clearly than in reference to the *Opus Postumum*. We have seen in the previous chapter that Erich Adickes was responsible for the eventual inclusion and publication of the *Opus Postumum* as part of the *Akademie-Edition* in 1936/1938, a century after Kant's death. His argument for inclusion was that the chronologically last Fascicle I was virtually complete by 1801 and thus at a time when Kant had still been in good (enough) health.[29] Following Adickes's intervention, the manuscript was no longer rejected as a whole. While Adickes observed early traces of Kant's decaying spirit in the manuscript from as early as 1799, it is the chronologically last Fascicle which Adickes reads as testament to Kant's increasingly detrimental senility. This senility is most visible, Adickes claims, in the fact that over more than two years, Kant only filled less than 120 print-pages, which is substantially less than his pre-senile pensum.[30] He further notes that one now encounters *anacoluthons* with increasing frequency, that is, unexpected syntactical discontinuities and thus logical and grammatical incoherencies of thought both within single sentences and between the sentences themselves, all of which are interpreted by Adickes as representative of Kant's diminishing mental powers. Adickes also notices that the late pages of the manuscript are filled with crossed-out words as well as crossed-out lines, which he reads as an indication that Kant now has trouble finding the proper word for what he wishes to express. From Folio 4 onwards, Adickes diagnoses that it has now become increasingly difficult for Kant to follow a continuous argument for longer than one folio page. Until finally, incapable of finishing full sentences, Adickes claims that Kant finds it difficult to think and write at all, taking recourse to abbreviations and finally single words.

> Without a doubt, [. . .] Kant lacks the necessary ability to conceive a larger whole not only in vague outlines but also to execute this plan step by step and expand it in detail. It would have required a certain minimum level of concentration and creative power [*Gestaltungskraft*], which Kant no longer possesses. [. . .] Even on favourable days, his mental and physical powers rapidly weaken during work, and instead of completing what he has started, he begins anew, only to suffer the same fate once again.[31]

Gerhard Lehmann also prominently refers to the *Opus Postumum* as the 'decisive document concerning Kant's age-related illness or so-called senility'.[32] Differentiating between Kant's age-related weakness and the onset of senility proper, which Lehmann dates to the year 1802–3 and refers to under the pejorative name of 'senile psychosis' (*Alterspsychose*),[33] he also notes the problematic status of Fascicle I in his 'Introduction' to the academy edition of the *Opus Postumum*. While critiquing Adickes at one moment for painting an image of Kant as quasi-schizophrenic in his violent distinction between Kant the person, the moral philosopher, the epistemologist or the transcendental philosopher,[34] Lehmann himself argues that, while the *Opus Postumum* as a whole must be read as containing important philosophical reflections proper and thus developments of Kant's critical philosophy, the chronologically last Fascicle I does contain *Senilia* proper. Changes in handwriting are interpreted by Lehmann as proof of Kant's increasing senility: a shaky hand, words that often need to be corrected, lines that are stroked through, with the lines themselves no longer appearing uniform and harmonious. Furthermore, he attests that the chronologically last Fascicle I is witness to increasingly mechanical-associative relations between representations in Kant's thought, sometimes of a playful and more often, however, of a compulsive nature.[35] Rather than inventing and thinking proper, Lehmann concludes that Kant found himself trapped in mechanical thought processes by the time he was writing Fascicle I.[36]

Schepelmann's refutation

In 2018, Maja Schepelmann published *Der senile Kant? Zur Widerlegung einer populären These* with the explicit attempt to refute the common claim that Kant had been befallen by senility towards the end of his life. To this end, Schepelmann engages in detailed and convincing historical and hermeneutic analyses to first show that Kant had anonymously published to a much larger extent during the last decade of his life than usually assumed, and second that what had hitherto been believed to be facts about Kant's physical and mental condition in fact derived from handed-down statements from a context that she characterizes as the circle of Kant's opponents and enemies, with the exception of letters between Kant and Rink.[37]

Next to a number of small anonymous publications, Schepelmann makes a convincing case to count two anonymously published books as in fact authored by Kant between 1798 and 1800. The first of the two is the 1798/1799 *Stimme eines Arktikers über Fichte und sein Verfahren gegen die Kantianer von D. K.*[38] Containing a 150-page-long refutation of Fichte, the book not only includes long citations of Fichte's philosophy but is also witness to the – previously thought to be anonymous – author's deep familiarity with Fichte's work. As such, Schepelmann's attribution of *Die Stimme des Arktikers* to Kant stands in direct opposition to Adickes's assessment, referred to earlier already, according to which Kant had supposedly lost the mental capacity to think himself into lines of thought foreign to his own from as early as 1791. Adickes claimed in his compendium to the *Opus Postumum* that Kant 'was so heavily engaged in the elaboration of his system and, on the other hand, due to advancing age, he had lost the capacity to immerse himself in foreign thought processes that significantly deviated from his own. Consequently, he did not attempt to intervene publicly in the development of his school with an authoritative statement [*Machtwort*]'.[39] Schepelmann's proposal to read *Die Stimme des Arktikers* as authored by Kant refutes both the idea that in the last decade of his life, Kant's mind had become too ponderous and recalcitrant to comprehend philosophical thought different from his own, as well as painting a picture of Kant as intimately familiar with and well engaged in the philosophical controversies of the time.

The second book attributed by Schepelmann to Kant is entitled *Sinnlichkeit und Verstand oder über die Prinzipien des menschlichen Wissens. Eine Kritik über Herrn Herders Metakritik von einem Wahrheitsfreunde*, anonymously published in 1800.[40] Schepelmann proposes that, following the weak and partly misguided answers by prominent Kantians[41] to attacks on Kantian thought in Herder's *Ideen, Kalligone,* but especially in the *Metakritik,*[42] Kant took to writing himself and anonymously published a 350-page-long refutation of Herder. According to Schepelmann, Kant had thus written nothing less than an additional 500 print-pages worth of philosophical commentary and criticism at a time in which he is frequently referred to as elderly and senile, and of which the official Kant corpus only cites publications of his prior lecture courses on logic, physical geography and pedagogy, edited and compiled by Jäsche and Rink.[43]

Schepelmann's second path of refuting the senility hypothesis traces and analyses the documentary evidence about Kant's supposed senility back to its historical origins. Schepelmann concludes that all documents invoked in order to prove Kant's senility derive from two contexts. The first historical context of documentary evidence regarding Kant's senility is nothing else than what Schepelmann calls the context of Kant's 'enemies and opponents'. Both Fichte, Schelling, Pörschke, and Beck suggested both already before as well as after Kant's death to varying degrees that imperfections in Kant's system are the product of Kant's elderly age, attempting to damage Kant's philosophical reputation by way of spreading rumours about his elderly condition and increasing senility.[44] The second historical context of documentary evidence regarding Kant's senility are letters by both Kant himself as well as his close companion Rink, who functioned as Kant's right hand from 1800 until the end of 1801.[45] References in Kant's own letters to his old age, his forgetfulness as well as weakening health should, however, according to Schepelmann, be interpreted as testament to Kant's rhetorical and self-ironic critique of his critics, who are spreading these rumours in the first place. Kant's self-ironic letters should thus not be read as evidence in support of his senility.[46] Regarding Rink's letters, Schepelmann points out that it is important to consider Rink's addressee when reporting of Kant's weakening mental and physical powers. Charles Francois Dominique de Villers reached out to Rink at various moments in an attempt to support himself financially in exile by selling autographs of Kant. Consequently, Rink's various references to Kant's declining physical and mental health in his letters to de Villers could and should be interpreted as Rink's (fatefully misguided) attempts to keep de Villers at a distance.[47]

Schepelmann does not deny that Kant had been afflicted by a 'virulent forgetfulness'[48] from an early age, followed by further physiological restrictions towards the end of his life. For Schepelmann, Kant was however able to resist the bodily phenomenon of forgetfulness with the strength of his will.[49] The manifold notes that Kant produced throughout his life, his later practice of jotting down notes into his *Memorienbüchlein* (Memory Booklet),[50] as well as the manifold biographical references contained in the *Opus Postumum*, should all be read as signs of Kant's intention of confronting his physical weakness by way of the strength of his will and understanding.[51] Furthermore, and on a methodological level, Schepelmann questions the practice employed across Kantian scholarship to interpret Kant's philosophical system and work as a whole by way of his biography, whereby conflating Kant the thinker with Kant the person. This method is first questionable, according to Schepelmann, because Kant had himself so rigorously attempted to separate his personal life from his philosophical one.[52] Second, however, Schepelmann shows that such cross-reading and conflation of Kant the person and philosopher was first performed and made acceptable by critics of Kant like Schopenhauer and Hegel.[53] The least one could do, Schepelmann seems to argue, is to problematize the philosophical as well as socio-political context in which these arguments first appeared and were made acceptable.

While Schepelmann clearly intends to rehabilitate Kant with her impressive refutation of the senility hypothesis, my aim in this final chapter is neither to further the senility hypothesis nor follow Schepelmann in her project of rehabilitating Kant and his late writing. While being intimately informed by both accounts, I propose to read philosophy's long history of referring to Kant's alleged senility, including the problematic reception history of the *Opus Postumum* problematically, in order to critically trace that and how this history is productive of a curious double discourse. This double discourse first appeared in the publication of Kant's first biographies in 1804, written by the Nicolovius-biographers, and has affected scholarly engagement with Kant's late writings and in particular the *Opus Postumum* until today.

Taking its cues from Borowski, Jachmann and Wasianski's assessments, the year immediately following Kant's death first puts forth a double discourse according to which Kant's last manuscript is not to be read philosophically as it is exemplary of a senile and thus pathological mind. Consequently, the manuscript should be read symptomatically as testament to Kant's decaying mind: his senility. As a consequence of this first symptomatic rather than critical and philosophical reception, the manuscript did not get included in the corpus of Kant's work. The manuscript is instead read as the product of Kant's forgetfulness and should itself be forgotten.

Following the first bifurcation of discourse on Kant's writings, further symptomatic readings are put forth by a variety of different authors in the decades following Kant's death, each shifting the line across which the discourse bifurcates into clinical and symptomatic versus critical and philosophical back and forth, sometimes as early as to the 1787 B-Edition of the *Critique of Pure Reason*. It was Adickes who succeeded in 1916 in internalizing the bifurcation line into the manuscript of the *Opus Postumum* itself, whereby the inclusion of the *Opus Postumum* into the published Kant edition became possible. As we have seen, it is only part of the chronologically last Fascicle I, which is to be read symptomatically according to both Adickes and Lehmann.[54]

As concerning Schepelmann, the previous section has introduced her refutation of the senility thesis *tout court*. At the same time, it appears that Schepelmann shifted the bifurcation line out of Kant's philosophical work in its entirety only to resurrect it and mark a difference between Kant's body and mind.

> The following can be presumed: Kant increasingly suffers from age-related, thus senile forgetfulness, but he can counteract it through aids [*Hilfsmittel*]. It would have had a much greater impact on his daily life than on the papers of the thinker. [. . .] It is appropriate to differentiate, in the last years, between the weak body and forgetful head [*Kopf*] on the one hand, and the strong will and power of thought of Kant on the other.[55]

By way of upholding a strong body/mind split when it comes to Kant, Schepelmann shifts the line, which she otherwise critiques and across which the discourse bifurcates, to enact a difference between Kant's physical life, which is here read symptomatically, and his intellectual, philosophical life of the mind, which is supposedly left unaffected by its physical counterpart and its pathologies.

The long and manifold attempts of Kant scholarship to distinguish between the healthy and the senile Kant are indicative of the threat that forgetting and forgetfulness pose to the universal project of Kant's philosophy. Kant's supposed forgetting threatens the integrity of his philosophical system and poses the difficult problem of how to differentiate between Kant's healthy, rational and universal writing on the one hand, and his pathological, irrational and symptomatic writing on the other.

At the same time, the long and manifold attempts of Kant scholarship to distinguish between the healthy and the senile Kant are also indicative of the problem that forgetting and forgetfulness pose to the universal project of philosophy in general. Throughout its history, philosophy has predominantly sought truth on the side of remembering rather than forgetting,[56] and in so doing juxtaposed the desired achievement of remembering against the common, unconscious misfortune of forgetting that can befall even the best of us. It is only in more recent modern times, as pointed out by Harald Weinrich, that philosophy has admitted forgetting as a possible object of philosophical inquiry and thus 'attempted to grant forgetting a certain truth as well'.[57] But when it comes to Kant, the previous section has shown that and how Kant scholarship has hitherto predominantly restricted itself to an inquiry into Kant's problematic forgetting of philosophy towards the end of his life, with none of the before-mentioned thinkers asking what Kant himself might have had to say about remembering and forgetting on the level of philosophical concepts themselves. In what remains of this chapter, I thus propose to take a different route and approach Kant's philosophy of forgetting from within the philosophical form of the double genitive, whereby both grammatical terms are called into question as both subject and object, respectively. What this means for the given conjuncture between philosophy, technics and forgetting, is to think forgetting at once from the perspective of the intersection of both Kant's philosophy of forgetting and his forgetting of philosophy. It is at this very intersection that technics will show itself once again.

From the art of memory to the art of forgetting

Memory, such is Stiegler's argument in his *Technics and Time* series, should not be mistaken as an anthropological or social constant. Rather, both anamnesic, embodied and thus internal, as well as external, technological and hypomnesic memory are subject to transformation through time, the study of the history of which, following Derrida and Sylvain Aroux, Stiegler calls grammatization.[58] Grammatization refers to the cultural and historical development of external, 'epiphylogenetic' memory systems,[59] which accumulate individual experiences and traces inscribed and collected in technical artefacts, thus enabling the preservation and dissemination of information beyond individual, genetic and epigenetic memory.

In the course of the historical development of external memory systems, the 1436 technical invention of the printing press, marking the beginning of the modern era that will eventually lead us to Kant, fundamentally changed 'natural', anamnesic memory. Elena Esposito's systems-theoretical essay *Die vergessenen Reste: Theorie und Praxis des blinden Flecks*[60] argues in this regard that the printed word, emancipated from its local context and addressed to an anonymous public, affords abstraction and stands in close proximity to forgetting.

> Memory based on texts is much stronger than the memory of the rhetorical tradition, which relies primarily on orality, because it leaves out much more data (it forgets much more), but allows for considering an incomparably greater variety of situations (it remembers much more). It is an abstract memory that relies on forgetting rather than directly on the capacity for remembering in order to enhance memory.[61]

Consequently, natural memory is claimed by Esposito to have become substantially weaker during the printing press era when compared to its pre-modern predecessors, since the key to memory now lies in the idea of order, provided by a metaphorical catalogue or archive, rather than in the laborious practice of metaphorically collecting data in the 'memory storage' of one's mind.[62] Consequently, one can observe a devaluation and eventual disappearance of earlier natural memory practices, including the ancient practice of the art of memory (*ars memoria*), in favour of a renewed and increased reflection and focus on forgetting with the beginning of the modern era set in motion by the invention of the printing press.[63]

Christine Abbt and Aleida Assmann also interpret philosophical modernity and, in particular, the subsequent enlightenment period as, above all, a renewed reflection on the meaning and role of forgetting. While in the mental spaces of the *ars memoria*, every possible object of thought could be given a place, only forgetting could not,[64] the enlightenment period would go on to elevate forgetting to a faculty (*Vermögen*) in its own right and thus articulate it as a condition of possibility for autonomous thought and judgment. Abbt shows that this can already be seen in the works of Michel de Montaigne, Francis Bacon, and René Descartes.[65] But nowhere can it be seen more than in Kant. According to Harald Weinrich, enlightenment philosophy, and in particular Kant's critical project, performed a turn away from memory in favour of forgetting.[66] 'Only by turning away [*Abwendung*] from memory, in the form in which it was still dominant in the academic philosophy that preceded him, could Kant become the "great destroyer in the realm of thought [*der große Zerstörer im Reiche der Gedanken*]" (Heine) and "all-crushing [*alles Zermalmende*]" (Mendelssohn).'[67] In opposition to scholastic, doctrinal philosophy and its accompanying focus on memory and the system of philosophy, Kant is thus read by Weinrich, Assmann, and Abbt as having put forth an image of enlightenment reason and the practice of philosophizing as aligned with forgetting, which is here however understood as a thoroughly rational work of forgetting in service of the autonomy of thought and the project of enlightenment rather than a mere physiological misfortune.

One can observe this rational practice of forgetting at work in the context of Kant's elaboration and explication of the philosophical process of abstraction in his *Anthropology from a Pragmatic Point of View*. Abstraction is here characterized as 'a real act of the cognitive faculty of stopping a representation of which I am conscious from being in connection with other representations in one consciousness'.[68] To abstract from something is thus in essence explicated by Kant as the capacity of turning away, both mentally and physically, from a representation in thought or actuality of which I am conscious. This action of turning away and abstracting, Kant goes on to claim, is a necessary condition for the autonomous use of the faculty of reason and should be frequently practised. 'To be able to abstract from a representation even when the senses force it on a person, is a far greater faculty than that of paying attention, because it demonstrates a freedom of the faculty of thought and the authority of the mind, *in having the object of one's representation under one's control* (*animus sui compos*).[69] As pointed out by Abbt, if abstraction needs to be understood as a condition for the autonomous use of reason, forgetting is itself the condition for abstraction,[70] since the unwanted representations need to be able to be let go of and forgotten for abstraction to properly function. Consequently, Kant's reflections on abstraction are interpreted by Abbt as testimony to the fact that Kant not only accepted forgetting as a condition of possibility of autonomous thought but that Kant was furthermore interested in exploring the question of what was to be forgotten as well as how.[71]

Steffen Häfner calls this Kantian notion of abstraction, reportedly perfected by Kant the person, a proto-version of what would eventually become known as thought-stopping (*Gedankenstopp*),[72] a psychotherapeutic technique to counteract mental rumination developed during the 1950s by cognitive psychology. There are a number of examples that one would do well to learn to abstract from, according to Kant. These include learning to overlook certain unpleasant facial features in the face of one's spouse, for instance, as well learning to abstract from missing buttons in your opposite, and so on.[73] There is, however, also a famous, late note written by Kant about his former butler Martin Lampe, which was found and reproduced for humanity by Wasianski in his biography and reads: 'The name Lampe must now be completely forgotten.'[74] To Wasianski, this late note about Lampe contains a curious contradiction and is read as a sign of Kant's weakness (*Schwäche*), since 'usually, one writes down what one does not want to forget'.[75] Kant's general, quasi-graphomanic practice of notetaking, particularly over the last years of his life, was, as we have seen, interpreted by the Nicolovius-biographers and broader Kantian scholarship as proof of Kant's senility and failing memory. For Schepelmann, Kant's notetaking practice was an expression of his strength of will in his project of overcoming his merely physiological forgetfulness by means of writing. According to Kant himself, the 'art of writing [*Schreibkunst*]' is indeed a fine substitute in case of a lack of memory, as he writes in the *Anthropology from a Pragmatic Point of View*.[76] To be 'safe with a notebook in the pocket is after all a great convenience, in order to recover precisely and without effort everything that has been stored in the head. The art of writing [. . .] takes place of the most extensive and reliable memory, and can [substitute] for its lack'.[77] While agreeing with Plato's *Phaedrus* on the one hand that '"the art of writing has ruined memory"',[78] Kant is at the same time far from advocating for a return to a dialectic and active engagement of the mind for the anamnesic recollection of knowledge on the other. He also does not propose to write to remember in the 'usual' sense. Employing the German word *ersetzen*, Kant proposes that the art of writing, and thus external, artificial memory, can substitute and thus supplement for the lack of 'internal memory'. The question, then, is whether Kant's late note concerning Lampe should be read less as an expression of Kant's forgetfulness and the dissolution of his prior mnemonic powers, as it might rather be a trace of Kant's desire to forget at will and of his active, conscious and long-practised rational technics of forgetting; a kind of *ars oblivionis* or *lethotechnics*.

The idea of such an ominous *art oblivionis* was first mentioned in various texts by Cicero, according to which an Athenian politician and general under the name Themistocles rejected the poet Simonides's offer to teach him his famous *mnemotechnics* (*ars memoria*). Themistocles, Cicero writes, wanted much rather to learn how to forget at will rather than to remember.[79] To this end, Kant polemically exclaimed in his *Anthropology from a Pragmatic Point of View* that 'there is no mnemonic art (*ars memoria*) in a sense of a general doctrine'.[80] As such, Kant categorically rules out the possibility of a mnemonic art in the context of his excursus on the faculty of memory (*Gedächtnis*) in the *Anthropology from a Pragmatic Point of View*. There are, he claims, three kinds of memorizing (*memorieren*). The first kind of memorizing is merely mechanical, and as the product of training, discipline and repetition, it barely possesses any intellectual importance for Kant.[81] The second kind of memorizing concerns the already mentioned mnemonic art, that is, the ancient 'method of impressing certain ideas on the memory by association with correlative ideas that in themselves have no relationship at all with each other'.[82] Objecting that the second kind of memory thus places a double burden on one's cognitive powers since in fact one needs to remember two rather than one idea, both of which are furthermore unrelated to one another, Kant introduces a third kind of memory, which he called judicious memorizing and which he claims to be the highest possible employment of memory and thus his preferred kind. Here, the faculty of memory works both under the rule of the power of judgment, which preselects what is to be remembered and what is to be abstracted from (forgotten), and the faculty of reason, which provides a system of topics into and according to which the preselected representations are to be sorted in a second step.[83] The third kind of memorizing proceeds along 'a table of the divisions of a system (for example, that of Linnaeus) where, if one should forget something, one can find it again through the enumeration of the parts that one has retained or else memorizing the sections of a whole made visible'.[84] One key to the third, 'good' kind of memory is thus indeed order over and above the laborious collection of data in one's 'memory storage'.[85] At the same time, however, the second key to the 'good' kind of memorizing is that, while enumerated as part and parcel of Kant's explication of memory, it is described as a method of 'finding' thoughts that one has already forgotten, that is, it sets off from the perspective of forgetting and employs the systematic form of a whole provided by reason as a way of remembering.

The Nicolovius-biographer Borowski reports that Kant not only practised judicious memorizing according to a table of divisions himself, but that he further advised his students to practice it as well.

> During his early years as a lecturer, he advised us, the students sitting around him, to consider the scientific knowledge we had gathered so far as divided into various compartments in our minds. And then, while reading a book or journal that presented a new idea previously unknown to us, always have the question at hand: into which category or compartment does this, which you are currently reading, belong – where do you place it? – This way, the read or newly learned material would make a more indelible impression on us; even if we were to forget the idea itself later on, we would always remind ourselves: Something related to this or that has been stored in this or that compartment – and with some effort, it would then present itself [*darstellen*] in its entirety. He believed that such categorization of what has been newly acquired also greatly contributes to the proper organization of our knowledge.[86]

Judicious memorizing according to the systematic topic of forms is both explained by Kant as beneficial in the retention of studied material and to find and thus remember what has been forgotten. Furthermore, such a systematic topic of forms employed in judicious memorizing was furthermore used by Kant in his process of writing and thus in the process of philosophizing.

The art of forgetting and the art of writing

Within the Kant corpus, the *Opus Postumum* occupies a unique position. As an unfinished manuscript, Lehmann contends that it should – properly speaking – neither be considered nor called an *Opus*. Rather, it should be read and studied 'as a series of drafts, in which Kant freely writes his thoughts into existence [*frei erschreibt*]'.[87] While unique in its unfinished state, this practice of freely writing his thoughts into existence was Kant's preferred method of philosophizing in general throughout his life, as attested to in his *Philippi Logik*, a set of lecture notes from an early logic seminar held by Kant during the summer semester of 1772, and referenced by Lehmann in the footnote to the before citation:

> At the beginning one meditates tumultuously. One must write out that which comes to mind, if an occasional thought comes to mind that one has never had in one's life. First, one jots down all thoughts, as one has them, without order. After that, one begins to coordinate and then to subordinate. If one wants to produce something, one must certainly complete the skeleton of the system in general, and subsequently divide this into chapters. Thus every elaboration [*Außarbeitung*] must proceed with three tasks [*Arbeiten*]:
>
> 1. One jots down all thoughts, without order.
> 2. One makes a general plan.
> 3. One elaborates [*arbeitet . . . aus*] all the parts.[88]

Adickes confirms that Kant thought 'with the pen in his hand'.[89] Conceding that he might have anticipated the general outline of his thought, the elaboration of the particulars are nevertheless said to have taken shape in writing. In reference to a number of posthumously found loose sheets, Adickes repeats that Kant 'as it was his habit, would clarify obscure points by writing down his reflections, thereby achieving clarity through the written fixation of his thoughts'.[90] According to Adickes, Lehmann and Kant himself, thinking *was* writing.

When it comes to the *Opus Postumum*, the many drafts of one and the same section confirm the essential role that writing and drafting played for the development of Kant's philosophical thought. Borowski described Kant's writing method in his Kant biography in the following way.

> He first made general outlines in his mind; then he worked on them in more detail. He wrote down on small pieces of paper what needed to be inserted here or there, or what needed closer explanation, which he simply added to that hastily written initial manuscript. After some time, he revised the whole thing

> once again and then transcribed it neatly and clearly, as he always did, for the printer.[91]

Further to this, the following remarkable observation was made by Vittorio Mathieu in his textual analysis of the *Opus Postumum*. Written mainly on large sheets of paper that, when folded, produce so-called folios of four pages each, Mathieu famously characterized the manuscript as 'cell-like (*zellenartig*)'.[92] Mathieu employed this Leibnizian image to point to the fact that any thought almost never extends beyond the boundary of its folio, or at least beyond the limit of its sheet.

> It is noteworthy that Kant's coherent text was never (with a single exception mentioned below) written beyond the boundaries of a single sheet, hardly ever extending beyond the limits of a single page. When space becomes limited, the handwriting is reduced in size, and when there is no space left at all, requiring some words to be written on the next page (such as the last three words of page 2 of the eighth sheet of the fifth section), these 'remaining words' are connected to their respective page by the symbol #. In principle, the unity of thought coincides with the given formal unity of the paper (one page, one sheet). In this way, the pages themselves become images of the thought process. Each individual sheet, and often each individual page, has the task of taking up the sequence of thought as a whole, thus assuming a synoptic function.[93]

According to Mathieu, Kant's philosophical process was determined by the material form of unity of his prosthetic milieu to the point of sharing the synthetic function of synopsis with the material form provided by the page. According to Kant's *Deduction* in the *Critique of Pure Reason*, synopsis was responsible for a first combination of the manifold of representation in the threefold synthesis. The unity of the page provides unity and systematicity to Kant's process of philosophizing, which, in the case of the *Opus Postumum*, concerned the process of thinking and writing the transition between metaphysics and physics. The fact that the overall form of unity of a folio sheet was then distinguished and organized according to four pages raises the question of the relation between the material form of unity provided by Kant's prosthetic milieu and his famous fourfold topic, the table of categories, employed at various stages throughout his work as an organizational schema. And indeed, in the chronologically early Fascicles, the headings of the folios are often entitled quantity, quality, relation and modality, and thus according to the four classes making up the table of categories laid out in the *Critique of Pure Reason*. This is the case, because in the chronologically early Fascicles, Kant intended to write/think the concept of matter according to the topic provided by the table of categories, as for instance in Fascicle IX, when Kant writes, 'the classification according to a priori principles can only be done in accordance with the system of categories. The motivating forces of matter, therefore, need to be considered based on their quantity, quality, relation, and modality'.[94] Following this first announcement of this distinction, Kant then proceeds to think the quantity of matter on sheet VII, page 1, followed by the quality of matter on sheet VII, page 2, after which relation and modality are left behind, only to begin the same process on the next sheet. The synoptic function provided by the page is thus more often than not coextensive with a topic, the thinking through of which is termed judicious memorizing in the context of the *Anthropology from a Pragmatic Point of View* and which can now be seen to also be intrinsic to Kant's practice of philosophizing, that is, of thinking as writing.

In a remarkable manner, Kant's approach to writing thus pays witness to the inadequacy of Stiegler's critique of Kant in his *Technics and Time* series, where he accused Kant of having been incapable of seeing, understanding, and thinking the understanding's 'veritable crutch'.[95] Stiegler's argument claimed that Kant failed to acknowledge that the unity of his own thought was only accessible to him through his books, notes and the manifold of other technical traces that surrounded him. The reason for this failure was argued by Stiegler to be the effect of Kant's transcendental idealist stance: due to Kant's insistence on the interiority of phenomena, he forgot (in the sense of repressed) his own prosthetic milieu as well as that of transcendental consciousness, the subject of his investigation. The preceding section has shown that Kant engaged in a practice of thinking *as* writing in which the synthetic function of synopsis, combining a manifold into a first unity, was shared with the unity provided by the material form and organization of the page. The manuscript of the *Opus Postumum* functions as an invaluable resource for studying the primordial role that Kant's prosthetic, technical milieu has always already played for the development and unity of his thought.

Engaging the relation of Kant and forgetting from the perspective of the double genitive that is a philosophy of forgetting, has finally shown that and how the practice of writing and externalizing might also have played a part in Kant's rational project of forgetting. Reading Kant's forgetting together with Weinrich, Assmann and Abbt has brought to the fore that and how Kant considered forgetting a condition of possibility of autonomous thought through the role it always already plays in the process of abstraction. Furthermore, Kant, the forgetful one, was shown to have participated in and furthered modernity's renegotiation of the role and responsibility of forgetting over and against remembering as a whole. Exposing the vital role that forgetting played within the enlightenment's larger project of critiquing established authorities, dogmas and philosophical systems, forgetting is engaged by Kant in service of, rather than in opposition to, the autonomy of thought and practical freedom on the level of the individual as well as the social. Kant, we must thus conclude, practised and advocated for a technics of forgetting in service of the autonomy of reason due to which his thought of technics essentially takes its cues from Themistocles rather than Simonides and thus from forgetting rather than remembering.

Notes

Introduction

1 Kant, Immanuel. *Opus Postumum*, trans. Eckart Förster, Michael E. Rosen (Cambridge: Cambridge University Press, 1993), 21:44. Hereafter OP.

> 'Titelblatt u Vorrede
> Die Welt als Universum
> In allen diesen Objecten ein Maximum Idee ergo unicum in allen 3 Fällen
>
> 1 Die theoretisch//speculative
> 2 Die technisch practische
> 3 Die moralisch practische Vernunft
>
> Aus Anschauungen, Begriffen a priori u Ideen
> Die Idee der Freiheit führt durch den categor. Imperativ auf Gott
>
> 1 die speculative
> 2 die practische
> 3 die technisch//practische
> 4 die moralisch//practische Vernunft in einem System.'

2 Lehmann, Gerhard. 'Die Technik der Natur', in *Beiträge zur Geschichte und Interpretation der Philosophie Kants* (Berlin: Walter de Gruyter & Co., 1969), 289.

3 Heidegger, Martin. *The Question Concerning Technology*, trans. William Lovitt (New York & London: Garland Publishing, 1977). This essay is derived from material that Heidegger first presented at his 1949 Bremen lecture entitled *Das Gestell*. The title is usually translated as *The Question Concerning Technology*, but in order to uphold the conceptual distinction between the German terms *Technik* and *Technologie*, which was fully in place at the time of Heidegger's essay, I will henceforth translate the German term *Technik* with the English neologism *technics*, which is also employed in the English translations of Gilbert Simondon's and Bernard Stiegler's works.

4 Simondon, Gilbert. *On the Mode of Existence of Technical Objects*, trans. Cecile Malaspina, John Rogove (Minneapolis: Univocal, 2017).

5 Ibid., 15.

6 Stiegler, Bernard. *Technics and Time, 1. The Fault of Epimetheus*, trans. Richard Beardsworth, George Collins (Stanford: Stanford University Press, 1998), ix (hereafter *TT1*).

7 The discussion between Kant and Schlosser was first brought to my attention by Catherine Malabou in Malabou, Catherine. *Before Tomorrow: Epigenesis and Rationality*, trans. Carolyn Shread (Cambridge: Polity Press, 2016), 6.

8 Schlosser, Johann Georg. *Plato's Briefe nebst einer historischen Einleitung und Anmerkungen* (Königsberg: Nikolovius, 1795), 182fn.

Part I

1 I will employ the terms 'technical thought', 'technics-thinking' as well as thinking of technics throughout this chapter, drawing on the German notion of *Technikdenken*, in order to highlight Heidegger, Simondon, and Stiegler's attempt to recover a post-metaphysical thought of technics within twentieth-century European philosophy.

Chapter 1

1 Heidegger, *The Question Concerning Technology*, 20.

2 Ibid., 4.

3 Heidegger, Martin. GA Band 76, *Vorstudien zum Technik-Vortrag* (Frankfurt a.M.: Klostermann, 2009), 347. 'Denken ist nicht *gegen* >die Technik< [. . .], sondern gegen die Oberflächlichkeit und Ahnungslosigkeit, mit der die Technik betrachtet wird.' My translation.

4 Heidegger, GA Band 76, 373:
'Die Technik ablehnen?
Ist um nichts weniger töricht als die Sonne ablehnen.
Was soll diese Ablehnung? – eine Selbsttäuschung!' My translation.

5 Heidegger, GA Band 76, 358. '*Die Frage nach der Technik*. Der Ton liegt auf der *Frage*. Sie zu entfalten ist vor allem nötig. Die Frage auszuhalten. Bedrängender als die Technik selbst ist die Frage nach ihr.' My translation.

6 Heidegger, *The Question Concerning Technology*, 6.

7 Ibid., 13–14.

8 Heidegger, GA Band 76, 401. The cover of the first Fascicle containing the manuscript is entitled *Technik 1940*.

9 Ibid., 309. '. . . Nachahmung der Handarbeit und des Naturvorganges.' My translation.

10 Ibid., 308. '. . . motorischen Charakter der wesentlich auf Krafterzeugung bezogen ist.' My translation.

11 Heidegger, *The Question Concerning Technology*, 14.

12 Ibid., 14.
13 Ibid., 16.
14 Ibid., 17.
15 Ibid.
16 Ibid., 19.
17 Ibid., 20.
18 Ibid., 4.
19 Heidegger, GA Band 76, 339. 'Das Wesen der Technik ist nichts Menschliches, aber die Technik gehört zum Menschen, weil der Mensch in das Wesen der Technik gehört. Je nach der Wesensgestalt der Technik [. . .] ist die Zugehörigkeit des Menschen in das Wesen verschieden und demgemäß die Art seines technischen Tuns anders.' My translation.
20 Heidegger, Martin. Band 90, *Zu Ernst Jünger* (Frankfurt a.M.: Klostermann, 2004), 287.
21 Jünger, Ernst. *The Worker: Dominion and Form,* trans. Bogdan Costea, L. P. Hemming (Evanston: Northwestern University Press, 2017), 97.
22 Heisenberg, Werner. 'Das Naturbild der heutigen Physik', in *Die Künste im technischen Zeitalter* (Hamburg: Rohwohlt, 1956), 42. 'Der Satz, daß der Mensch nur noch sich selbst gegenüberstehe, gilt aber im Zeitalter der Technik noch in einem viel weiteren Sinne.' My translation.
23 Heidegger, GA Band 90, 297. 'Wer ist der Mensch – der sich da selbst begegnet?' and a few lines later continuing 'Der *Mensch* ob so etwas überhaupt möglich? Und *wenn* für möglich gehalten – ob nicht in dieser Meinung die größte Verblendung.' My translation.
24 Heidegger, *The Question Concerning Technology*, 27.
25 Ibid., 26.
26 Ibid., 27.
27 See Heidegger, GA Band 9, 339.
28 Heidegger, Band 90.
29 Heidegger, Band 90, 217. See also 218: 'Jünger hatte nicht das *Buch* mit dem Titel >>Nietzsche, Der Wille zur Macht << im Tornister – sondern er wurde von Feuer und Blut, von Tod und der Arbeit, vom Schweigen und Donnern der Materialschlacht *als Erscheinungen des Willens zur Macht betroffen*.' 'Jünger did not only carry the book *Nietzsche, The Will to Power* in his knapsack – but he was affected by the fire and blood, by death and work, by the silence and thunder of the battle of materiel *as appearances of the will to power*.' My translation.
30 Heidegger, GA Band 90, 214. '. . . faßt in der Wirklichkeit Fuß, die Nietzsche denkerisch erlitten hat.' My translation.
31 Ibid., 218. '. . . die Stärke und Entschiedenheit der Besinnung und des Wortes.' My translation.
32 Heidegger, GA Band 90, 37. '. . . geschichtlichen Grunde des Wesens des vollendeten Zeitalters der Neuzeit.' My translation.

33 Ibid.

34 Ibid., 214.

35 Ibid., 215.

36 Ibid., 255. ‘. . . die Beschreibung schreibt uns selbst ein in die Zugehörigkeit zu diesem Wirklichen.’ My translation.

37 Ibid., 265.

38 Ibid., 227. ‘Nirgends zergliedert und beschreibt dieser denkende Krieger nur eine vorhandene geschichtliche <<Situation>>; sein Denken selbst ist eine Gestalt des Willens zur Macht – ; in Jüngers Sprache: das Denken hat Arbeitscharakter.’ My translation.

39 Ibid., 37. ‘. . . Erst wenn wir aus der Grundstellung denken und sie selbst nochmal als Geschichte (des Seyns) begriffen haben, wird entscheidbar, inwiefern und ob überhaupt Jüngers <<Werk>> nur das Wunschgebilde eines in bestimmter Weise veranlagten <<Individuums>> ist, oder ob es durchaus auf dem geschichtlichen Grunde des Wesens des vollendeten Zeitalters der Neuzeit aufruht und dieses zum Wort bringt.’ My translation.

40 Ibid., 213. ‘Und dennoch – über all dieses hinweg vollzieht er doch eine Hinführung in das Wirkliche des Willens zur Macht und macht rücksichtslos ernst mit diesem Wirklichen.’ My translation.

41 Ibid., 258–9. ‘Und es muß sich zeigen und kann sich freilich nur einem Fragen und die der bloßen Beschreibung offenbaren, daß das <<Wirkliche>>, das Jünger im Lichte der Nietzscheschen Metaphysik erfährt, keine neue Wirklichkeit ist und nicht das Anzeichen des Übergangs in ein neues Zeitalter, sondern diese Wirklichkeit ist nur die äußerste Vollendung und schrankenlose Ausbreitung und Entfesselung dessen, was wir als die Wirklichkeit der schon seit 300 Jahren währenden <<Neuzeit>> zu erkennen haben.’ My translation.

42 Ibid., 282.

43 Heidegger, GA Band 76, 339. ‘Wenn wir es vermögen, das Wesen der Technik sachgemäß auch nur zu erfragen, wandelt sich dabei das Fragen in das erörternde Sagen, wandelt sich das Denken. Wir bewegen uns nicht mehr im Vorstellen von Gegenständen, die es auch gibt und die jetzt eine besondere Herrschaft ausüben. Wir gelangen unterwegs – wohin? Solches Denken ist nicht mehr stellbar *im* Ge-stell und durch dieses – es sei denn zufolge einer stets möglichen metaphysischen Umdeutung.’ My translation.

44 In his later book *An der Zeitmauer*, Jünger returns to and develops the themes of the worker further, as he becomes more critical of the age of the worker and moves closer to Heidegger’s position. See Jünger, Ernst. *An der Zeitmauer* (Stuttgart: Klett-Cotta, 2013).

45 Heidegger, *The Question Concerning Technology*, 22.

46 Ibid.

47 Ibid.
48 Ibid., 23.
49 Heidegger, GA Band 76, 299.
50 Ibid., 310.
51 Ibid., 294. '. . . die eigentliche Vollendung der >Metaphysik< . . .' My translation.
52 Ibid., 285.
53 Ibid., 296. '. . . Wahrheit, in der das Seiende steht (dieses Stehen als Seynsgeschick *selbst*) und nach dem Seyn, das das Seiende in solche Wahrheit losgelassen hat.' My translation with the help of Cooper Francis. Heidegger here uses the archaic 'Seyn' from the more common 'Sein' found in Being and Time (Sein und Zeit) in order to evoke a sense of the forgotten or concealed dimensions of Being that have been obscured by the development of technology and the modern world.
54 See also Stiegler, *Technics and Time 1*, 5.
55 Heidegger, GA Band 90, 80.
56 Ibid., 152. '. . . Grundlegung des neuzeitlichen Menschentums.' My translation.
57 Martin Heidegger. *The Question Concerning the Thing*, trans. James D. Reid, Benjamin D. Crowe (London: Rowman & Littlefield, 2018), 72.
58 Ibid., 52.
59 Ibid., 38.
60 Ibid., 50.
61 Ibid., 145.
62 Heidegger, *The Question Concerning Technology*, 21.
63 Ibid.
64 Ibid., 90.
65 Kant, Immanuel. *Critique of Pure Reason*, trans. Paul Guyer, Allen W. Wood (Cambridge: Cambridge University Press, 1999), A247/B304. (Hereafter CPR).
66 CPR B163.
67 Except for beautiful and living things described in Kant's 1790 *Critique of the Power of Judgment*.
68 Heidegger, *The Question Concerning the Thing*, 38.
69 Lehmann, 'Die Technik der Natur', 293.
70 There are, however, plenty of references to the *Opus Postumum* in Heidegger's lecture course on Schelling and German Idealism (1941–3). See Heidegger, Martin. *GA Band 86: Hinweise und Aufzeichnungen, Seminare Hegel – Schelling* (Frankfurt a.M.: Klostermann, 2011), 246; Heidegger, *The Question Concerning the Thing*, 53.
71 See Hui, Yuk. *Recursivity and Contingency* (Washington, DC: Rowman&Littlefield Publishers, 2019).

Chapter 2

1 Simondon, Gilbert. *On the Mode of Existence of Technical Objects.* trans. Cecile Malaspina, John Rogove (Minneapolis: Univocal, 2017).

2 Simondon's twofold doctorate remained the only published work during his lifetime. His minor thesis *Du mode d'existence des objets techniques* (*On the Mode of Existence of Technical Objects*) was published in its entirety in 1958, while at first only the first half of his major thesis *L'Individuation à la lumiére des notions de forme et d'information* (Individuation in light of notions of form and information) was published in 1964. Following his death in 1989, the second part of his major thesis would finally follow. It was only in 2005 that the French publisher Jerome Millon would publish a complete edition of his main thesis, containing both parts. While a partial, unofficial English translation of *On the Mode of Existence of Technical Objects*, undertaken by Ninian Mellamphy, has been circulating since the late 1980s, it was not until 2017 that the first complete English translation undertaken by Cecile Malaspina and John Rogove was published by Univocal. In 2022, Simondon's major thesis was published in English by University of Minnesota Press in a translation undertaken by Taylor Adkins.

3 Heidegger, Martin. *Qu'est-ce que la métaphysique?*, trans. Henry Corbin (Paris: Gallimard, 1938).

4 Heidegger, *The Question Concerning Technology*, 20.

5 Simondon, *On the Mode of Existence of Technical Objects*, 176.

6 Ibid., 170.

7 Simondon, Gilbert. *L'Individu et sa genèse physico-biologique* (Paris: Presses universitaires de France, 1964).

8 Deleuze, Gilles. *Desert Islands and Other Texts 1953–1974*, trans. Michael Taormina (New York: Semiotext(e), 2004), 88.

9 Simondon, Gilbert. *Individuation in Light of Notions of Form and Information,* trans. Taylor Adkins (Minnesota: University of Minnesota Press, 2020), 4.

10 CPR A247/B304.

11 Simondon, *Individuation in Light of Notions of Form and Information*, 4.

12 Ibid., 1.

13 Ibid.

14 Ibid.

15 Ibid., 3.

16 Simondon, *On the Mode of Existence of Technical Objects*, 242.

17 Ibid.

18 Simondon, *Individuation in Light of Notions of Form and Information,* 3.

19 Simondon, *On the Mode of Existence of Technical Objects*, 248.

20 Ibid., 249.

21 Ibid., 248.
22 Simondon, *Individuation in Light of Notions of Form and Information,* 36.
23 Ibid.
24 Simondon, *On the Mode of Existence of Technical Objects*, 225.
25 Ibid., 249.
26 Ibid.
27 Ibid.
28 Ibid., xiii.
29 Ibid., 15.
30 Ibid., 141. 'One could use the term "autocratic philosophy of technics" for a philosophy that takes the technical ensemble as a place where machines are used in order to obtain power [*puissance*]. The machine is only a means; the end is the conquest of nature, the domestication of natural forces by means of a first act of enslavement: the machine is a slave whose purpose is to make other slaves. Such a dominating and enslaving inspiration can coincide with the quest for man's freedom. But it is difficult to free oneself by transferring slavery onto other beings, men, animals, or machines; to reign over a people of machines that enslave the entire world is still to reign, and every reign presupposes the acceptance of the schemas of enslavement.'
31 Ibid., 225.
32 Combes, Muriel. *Gilbert Simondon and the Philosophy of the Transindividual*, trans. Thomas LaMarree (Cambridge: MIT Press, 2013), 7.
33 Simondon, *Individuation in Light of Notions of Form and Information,* 7.
34 Kant, Immanuel. *Critique of Practical Reason*. In *Practical Philosophy*, trans. M. J. Gregor (Cambridge: Cambridge University Press, 1996), 8:35.
35 Kant, *Kant's Practical Philosophy,* 8:36.
36 Simondon, *On the Mode of Existence of Technical Objects*, 103.
37 Ibid., 106.
38 Ibid., 103.
39 Ibid., 110.
40 Ibid., 109.
41 Ibid., 110.
42 Ibid., 113.
43 Ibid.
44 Ibid.
45 Ibid., 110.
46 Ibid.
47 Ibid., 111.
48 Ibid., 120.
49 Ibid., 117.

50 Ibid., 15.
51 Ibid.
52 Ibid., 118.
53 Ibid., xiii.
54 Ibid.
55 Ibid., 85.
56 Ibid., 111.

Chapter 3

1 Stiegler, *TT1*, ix.
2 Derrida, Jacques. *Of Grammatology*, trans. Gayatri Chakravorty Spivak, corrected edition (Baltimore: Johns Hopkins University Press, 1997), 3–8 and 25–35. In his seminal work *Of Grammatology*, Jacques Derrida argues that philosophy and Western thought in general, is founded on the repression of writing and the privileging of speech. According to Derrida, the history of Western philosophy has been characterized by a pervasive logocentrism, which posits speech as the primary mode of communication and knowledge, and writing as a mere representation of speech. This privileging of speech over writing, Derrida argues, is an act of repression, as it overlooks the essential role that writing plays in the production and dissemination of knowledge.
3 Stiegler, *TT1*, ix.
4 Stiegler, Bernard. *Technics and Time, 3. Cinematic Time and the Question of Malaise*, trans. Stephen Barker (Stanford: Stanford University Press, 2011), 176–7. (Hereafter *TT3*).
5 Stiegler, *TT1*, 6.
6 Espinoza, Tania. 'The Technical Object of Psychoanalysis', in *Stiegler and Technics*, ed. Christina Howells, Gerald Moore (Edinburg: Edinburgh University Press, 2013), 151.
7 Stiegler, *TT1*, xii.
8 Ibid., 25.
9 Ibid., 114.
10 Vernant, Jean-Pierre. *Myth and Society in Ancient Greece*, trans. Janet Lloyd (New York: Zone Books, 1990), 183–202.
11 Stiegler, *TT1*, 193.
12 Ibid., 152.
13 Ibid., 140.
14 Ibid., 155.
15 Ibid., 16.

16 Ibid., 244.
17 Heidegger, Martin. *Sein und Zeit* (Tübingen: Max Niemeyer Verlag, 2006).
18 Stiegler, *TT1*, 235.
19 *Ibid.*
20 Stiegler, *TT3*, 141.
21 Stiegler, *TT1*, 193.
22 Ibid., 194.
23 Ibid., 2.
24 Ibid., 17.
25 Ibid., 49.
26 Ibid., 1.
27 Ibid., 206.
28 Stiegler, *TT3*, 197.
29 Ibid., 145.
30 Ibid., 92.
31 See also Prigogine, Ilya, Stengers, Isabelle. 'Science and Change', in *Order out of Chaos*, (London: Verso Books, 2017), xiii.
32 Stiegler, *TT3*, 188.
33 As discussed in Chapter 2, Simondon argues in *On the Mode of Existence of Technical Objects* that insight into technical operation proper could remain structurally unquestioned until the invention of post-artisanal technical objects.
34 Stiegler, *TT3*, 188.
35 Stiegler, *TT1*, 2.
36 For Stiegler, Karl Marx's *Grundrisse* was the first to begin work on the theory of the evolution of technics. See Stiegler, *TT1*, 2.
37 Stiegler, *TT3*, 172.
38 Ibid., 21.
39 Ibid.
40 Ibid.
41 Ibid., 67.
42 Ibid.
43 Ibid., 188.
44 Ibid., 193.
45 Ibid., 189.
46 Vernant, Jean-Pierre. *Myth and Thought among the Greeks*, trans. Janet Lloyd, Jeff Fort (New York: Zone Books, 2006), 306. Vernant points out that for the ancient Greeks, the process of becoming and change lacked the essential scientific criterion of *akribeia* (precision) necessary for the scientific *episteme*. This is explained by Vernant to be the case because there was no precise way of measuring time yet.

47 In both *TT1*, 2 and Stiegler, Bernard. 'Pharmacology of Spirit', in *Theory after Theory*, ed. Jane Elliott, Derek Attridge (London: Routledge, 2011), 295, Stiegler employs the ancient Greek term *doxa* in line with the Platonic distinction between *episteme* and *doxa*, that is, knowledge and belief or opinion.
48 See, for instance, Moss, Jessica. *Plato's Epistemology: Being and Seeming* (Oxford: Oxford University Press, 2021), 151, who translates from Plato, *Sophist 234c-d* '[The sophist can] bewitch people who are young and standing still further from the truth of things, showing them spoken images about everything, (a) so as to make the things spoken *seem* true, and the speaker [seem] to be the wisest of all about everything . . . [But when later the students] grasp clearly the things that *are*, they will (b) change their former *doxai*'.
49 Stiegler, *TT3*, xii.
50 Ibid.
51 Ibid., xi.
52 Ibid., 1.
53 Ibid., 141.
54 Ibid., 144.
55 Ibid., 49.
56 Ibid., 48.
57 Adorno, Theodor W., Horkheimer, Max. *Dialectic of Enlightenment*, trans. John Cumming (New York: Continuum, 1989), 124.
58 Kant, CPR A141/B181.
59 Stiegler, *TT3*, 42.
60 Ibid., 55.
61 Ibid., 41.
62 Ibid., 191.
63 Ibid.
64 Kant, Immanuel. 'On the Common Saying: That May Be Correct in Theory, But Is of No Use in Practice', in *Practical Philosophy*, trans. Mary Gregor (Cambridge: Cambridge University Press, 1996), 277. 'Now if an empirical engineer tried to disparage general mechanics, or an artilleryman the mathematical doctrine of ballistics, by saying that whereas the art of it is nicely thought out it is not valid in practice since, when it comes to the application, experience yields quite different results from theory, one would merely laugh at him (for, if the theory of friction were added to the first and the theory of the resistance of the air to the second, hence if only still more theory were added, these would accord very well with experience).'
65 Stiegler, *TT3*, xii.
66 Sigmund Freud, *Totem an Taboo*, trans. J. Strachey (London: Routledge Classics, 2004), 35.

67 Kant, CPR A141/B181.
68 Stiegler, *TT3*, 176–7.

Chapter 4

1 Stiegler, *TT3*, 180.

Part II

1 CPR, Aviii.
2 Ibid.
3 Ibid., Avii. Guyer and Woods here translate *Vermögen* as capacity, but for terminological precision I propose translating it as faculty.
4 CPR A5/B8.
5 CPR Aviii.

Chapter 5

1 Katsafanas, Paul. 'The Emergence of the Drive Concept and the Collapse of the Animal/Human Divide', in *Oxford Philosophical Concepts: Animals*, ed. P. Adamson, G. Fay Edwards (Oxford: Oxford University Press, 2018), 239.
2 See *Deutsches Wörterbuch von Jacob Grimm und Wilhelm Grimm*: <https://www.dwds.de/wb/dwb/trieb> . Translation into English from *Dictionary of Untranslatables*, entry on *Drive* written by Alexandre Abensour, ed. Barbara Cassin, trans. Steven Rendall, Christian Hubert, Jeffrey Mehlman, Nathaniel Stein, Michael Syrotinski (Princeton: Princeton University Press, 2014), 231.
3 Ibid.
4 The concept of *Trieb* will be granted a renewed, explicit treatment in Kant's 1790 *Critique of the Power of Judgment* in a discussion of Johann Friedrich Blumenbach's *Bildungstrieb* (formative drive) and thus a decade after the publication of the 1781 A-Edition of the *Critique of Pure Reason* where reason's drive first features. For methodological reasons, I will not take recourse to Kant's later adoption and reformulation of the Blumenbachian *Bildungstrieb* for the discussion of drive (*Trieb*) in the *Critique of Pure Reason*.
5 CPR A15/B29.
6 CPR B21–22.

7 CPR A710/B738.
8 CPR A797/B825.
9 CPR B22.
10 CPR A5/B8.
11 CPR B22.
12 CPR Avii. Once again I propose to translate *Vermögen* as faculty.
13 CPR Axii.
14 CPR B7. These, then, are the questions of ontology, psychology, theology and cosmology.
15 CPR Aviii.
16 CPR A61/B86.
17 The six logical works comprising Aristotle's *Organon* are *Categories, On Interpretation, Priori Analytics, Posterior Analytics, Topics* and *On Sophistical Refutations.*
18 Lambert, Johann Heinrich. *Neues Organon oder Gedanken über die Erforschung und Bezeichnung des wahren und dessen Unterscheidung vom Irrtum und Schein* (Berlin, Boston: Akademie Verlag, 1990).
19 Bacon, Francis. *Neues Organon Teilband 1.* Lateinisch – deutsch (Hamburg: Felix Meiner Verlag, 1990).
20 See Weatherby, Leif. *Transplanting the Metaphysical Organ* (New York: Fordham University Press, 2016), 22.
21 Diogenes Laertius. *Lives of Eminent Philosophers* (Cambridge: Cambridge University Press, 2017). Book 10: 28–30.
22 Ibid., Book 10:31–2. What is important here is that the Epicurean criteria for truth 'are our sensations, preconceptions, and feelings'. We will see how Kant, in Epicurean fashion, revalues the sensible in the *Transcendental Aesthetic* by claiming that sensibility is not merely a confused version of the rational, but rather that the only way objects can be given to us immediately is in sensibility.
23 Kant taught logic lectures for about forty years throughout his life. In 1800, G. B. Jäsche compiled and published the *Jäsche Logik* at Kant's request, compiled from Kant's lecture notes as well as the notes from students. For the difficulties of working with the text, see Boswell, Terry. 'On the – Textual Authenticity of Kant's Logic', *History and Philosophy of Logic* 9, no. 2 (1988): 193–203.
24 Kant, Immanuel. *Lectures on Logic*, trans. J. Michael Young (Cambridge: Cambridge University Press, 1992), AA 9:14.
25 AA 9:14.
26 CPR A795/B823.
27 CPR A11/B25.
28 Santozki, Ulrike. *Die Bedeutung antiker Theorien für die Genese und Systematik von Kants Philosophie Eine Analyse der drei Kritiken* (DeGruyter: Berlin, 2006), 74.

29 CPR Avii.

30 Deleuze, Gilles. 'What Can a Body Do?', in *Expressionism in Philosophy: Spinoza* (New York: Zone Books, 1992), 217–35.

31 Caygill, Howard. *A Kant Dictionary* (Oxford: Blackwell Publishing, 1995), 190, entry on faculty (*Vermögen*).

32 Baumgarten, Alexander. *Metaphysics. A Critical Translation with Kant's Ellucidations, Selected Notes and Related Materials*, trans. Courtney Fugate, John Hymers (London: Bloomsbury Academic Publishing, 2018), §220.

33 This is an important distinction that Kant employed in the overall structure of the *Critique of Pure Reason*. While the first half of the *Doctrine of Element*s engages the faculties of sensibility and understanding, the second half deals with the power of judgment and thus the actualization of the conditions of possibilities.

34 Kant, Immanuel. 'The only Possible Argument in Support of a Demonstration of the Existence of God', in *Theoretical Philosophy, 1755–1770,* trans. David Walford, Ralf Meerbote (Cambridge: Cambridge University Press, 2007), 2:72.

35 CPR B100.

36 CPR A492/B521.

37 CPR A247/B303.

38 CPR A796/B824.

39 CPR B188.

40 CPR A62/B87.

41 CPR A63/B88.

42 CPR A796/B825.

43 CPR A55/B80.

44 CPR A19/B33.

45 In the *Appendix,* Kant will reaffirm that the relation of cognition to an object is only enabled by sensibility, which is called another tool (*Werkzeug*). 'For in that lies the mystery of the origin of our sensibility. Its relation to an object, and what might be the transcendental ground of this unity, undoubtedly lie too deeply hidden for us, who know even ourselves only through inner sense, thus as appearance, to be able to use such an unsuitable tool of investigation to find out anything except always more appearances, even thought we would gladly investigate their non-sensible cause.' A278/B334.

46 If taken together, the a priori forms of intuition and understanding make up what Kant calls conditions of experience, since, rather than coming to us from the experience of any given particular object, they are what first makes experience possible.

47 CPR A20/B34.

48 Ibid.

49 Heidegger, *The Question Concerning Technology,* 21.

50 Ibid.

51 CPR A247/B304.

52 CPR Axii.

53 See Chapter 1.

54 Malabou, *Before Tomorrow: Epigenesis and Rationality*, 6.

55 Schlosser, *Plato's Briefe nebst einer historischen Einleitung und Anmerkungen*, 182fn.

56 AA 8:404. 'Die wegwerfende Art über das Formale in unserer Erkenntnis (welches doch das hauptsächliche Geschäft der Philosophie ist) als eine Pedanterei, unter dem Namen einer "Formgebungsmanufaktur" abzusprechen, [. . .]'. After careful consideration, I have come to the decision of providing my own translation of this line from Kant's 'On a recently prominent tone of superiority in philosophy', despite there being two English translations available, both by G. Hatfield, M. Friedman, H. Allison and P. Heath in the Cambridge University Press edition, and by Peter Fenves in the Johns Hopkins University Press edition. Both translations take liberty in their terminology and interpretation of Kant's response to Schlosser, which in my understanding obscures comprehension and could lead to misinterpretations.

57 Ibid., 'In beiden Feldern (des Theoretischen und Praktischen) ist es nicht eine plan-oder gar fabrikenmäßig (zum Behuf des Staats) eingerichtete willkürliche Formgebung, sondern eine vor aller das gegebene Object handhabende Manufactur, ja ohne einen Gedanken daran vorhergehende fleißige und sorgsame Arbeit des Subjects, sein eigenes (der Vernunft) Vermögen aufzunehmen und zu würdigen'; In Kant, Immanuel. *Theoretical Philosophy after 1781*, trans. G. Hatfield, M. Friedman, H. Allison, P. Heath (Cambridge: Cambridge University Press, 2010), 443, the German original is translated as 'In both fields (theoretical and practical) it is not an arbitrary *form-giving* undertaken *by design*, or even *machine-made* (on behalf of the state), but above all a *piece of handwork*, dealing with the given object, and indeed with no thought of taking up and evaluating the preceding industrious and careful work of the subject, his own faculty (of reason)'. Not only is the notion of machine introduced in the translation, thus setting up an opposition between machine and handwork that is absent from the German original, but Kant is also now affirming the manufacture of dealing with the given object and denying the careful work of the subject. Furthermore, the terminology of 'manufacture', which Kant takes from Schlosser's accusation, is not retained, obscuring the difference between Kant's own position and that which he is trying to distance himself from. The original meaning is inverted. In the 1998 *Raising the Tone of Philosophy: Late Essays by Immanuel Kant, Transformative Critique by Jacques Derrida* (Baltimore: Johns Hopkins University, 1993), 70, Peter Fenves translates the same quote as 'In both fields (the theoretical, the practical) there is no arbitrary form-giving set up (on the behalf of the State) in accordance with a plan or an industry, but there

is a manufacture that handles, above all, the object given; and indeed, without a preconceived thought, there is an ever proceeding labor and careful work of the subject to takeup and appraise its own faculty (of reason)'. This translation is closer to the original, but the notion of industry is introduced, and the manufacture of dealing with the given object is affirmed in its difference to the 'labour and careful work', which Kant only refers to with the simple German *Arbeit*. Due to these translation difficulties, I have chosen to use my own translation in the discussion of this important reference.

58 Kant, 8:404.

59 This can also be seen in Kant's reference to 'fabrikenmäßige Arbeit' in his *Lectures on Anthropology*. See AA 7:148.

60 AA 8:404.

61 Schäfer, Dieter. 'Aspekte der Wirtschaftsgeschichte Würzburgs vom Ausgang des Alten Reichs bis zur Gegenwart. Probleme, Projekte, Entwicklungen, Märkte, Betriebe, Firmen, Niederlassungen, Beschäftigung, Unternehmer und die Rolle der Stadt in zwei Jahrhunderten', in *Geschichte der Stadt Würzburg*, ed. Ulrich Wagner (Darmstadt: Wissenschaftliche Buchgesellschaft, 2001), 1319n48.

62 AA 8:404. 'In beiden Feldern (des Theoretischen und Praktischen) ist es nicht eine plan-oder gar fabrikenmäßig (zum Behuf des Staats) eingerichtete willkürliche Formgebung, sondern eine vor aller das gegebene Object handhabende Manufactur, ja ohne einen Gedanken daran vorhergehende fleißige und sorgsame Arbeit des Subjects, sein eigenes (der Vernunft) Vermögen aufzunehmen und zu würdigen.'

63 Ibid.

64 See Malabou, *Before Tomorrow: Epigenesis and Rationality*, 6.

65 CPR A77/B103.

66 CPR A15/B29.

67 CPR A138/B177. Whereas the original German text speaks of the emerging schemata as 'ein Drittes', the English translation by Paul Guyer and Allen Wood in the Cambridge University Press edition translate this 'third' as a 'third thing'. For sake of precision and clarity and in order to avoid ontologizing tendencies, I will in the following refer to the Kantian schemata as thirds, leaving their ontological status undetermined for the time being.

68 Kant, Immanuel. *Critique of the Power of Judgment*, trans. Paul Guyer, Eric Matthews (Cambridge: Cambridge University Press. 2000). First Introduction §V 212. Hereafter CJ.

69 CPR A51.

70 CPR A141/B181.

71 Ibid.

72 CPR A61/B85.

73 Ibid.

74 CPR A135/B174.
75 CPR A141/B181.
76 CPR A626/B654.
77 CPR A832/B860.
78 See also Seibicke, Wilfried. *Technik. Versuch einer Geschichte der Wortfamilie um τέχνη in Deutschland vom 16. Jahrhundert bis etwa 1830* (Düsseldorf: VDI Verlag, 1968), 184.
79 Aristotle. *The Nicomachean Ethics*, trans. J. A. K. Thomson (London: Penguin Books, 2004), 1139b23.
80 CJ 5:303.
81 CPR A11/B25.
82 On Kant's elaboration of the aporia of judgment, see Caygill, Howard. *Art of Judgment* (Oxford: Basil Blackwell, 1989), 4–5. The aporia of judgment designates the infinite need for ever further rules to be able to distinguish whether something stands under a rule or not. In his aim of finally 'solving' the aporia, Kant's account of the exercise of judgment, however, remains circular. Caygill shows how judgment 'is found to always presuppose judgment: its discriminations require rules, and its rules require discriminations'. The same circular problem also applies to transcendental logic.
83 CPR A141/B181.
84 Ibid.
85 CJ 5:303.

Chapter 6

1 CPR A707/B735.
2 Kant, *Lectures on Logic*, 9:18.
3 Ibid.
4 Seibicke, Wilfried. Technik. Versuch einer Geschichte der Wortfamilie um τέχνη in Deutschland vom 16. Jahrhundert bis etwa 1830. Düsseldorf: VDI Verlag, 1968.
5 Ibid., 6.
6 Ibid., 7.
7 Ibid., 148.
8 Ibid., 134.
9 Sebestik, Jan. 'The Rise of the Technological Science', *History and Technology: An International Journal* (1983): Vol. 1:1, 29.
10 Ibid., 31.
11 Seibicke, *Technik*, 101.
12 Ibid., 102.
13 Ibid., 102–3.

14 Micraelius, Joh. *Joh. Micraelii lexicon philosophicum* (Jenae: Impensis J. Mamphrasii, 1653), Sp. 1058. '[. . .] doctrina de proprietatibus, deque ordine & numero disciplinarium, sive illae sint Philologicae, ut Grammaticae, Rethoricae &c. sive Philosophicae, tam theoreticae quam practicae; sive superiorum facultatum, ut Theologicae, Juridicae & Medicae.' My translation.

15 Walch, Johann G. *Philosophisches Lexicon* (Leipzig: Verlegts Joh. Friedrich Gleditschens seel. Sohn, Johann G., 1733), Sp. 2514f. 'Technologie heißt die Lehre von den Kunst-Wörtern, wodurch man insgemein solche Wörter verstehet, welche Sachen bedeuten, die in einem gewissen Stand sich befindlichen Personen eigen sind. [. . .] Man kann sie in ungelehrte und gelehrte Teilen, davon iene in den Handwercken und gemeinen Künsten; diese in den gelehrten Handwercken und gemeinen Künsten; diese in den gelehrten Wissenschaften vorkommen, und wieder entweder allgemeine oder besondere sind; Die allgemeinen welche von allen Gelehrten gebraucht werden, befinden sich in der Philosophie, sonderlich in der sogenannten Ontologie, z.E. causa, effectus, principium, principatum, subiectum, adiunctum, substantia, accidens u.d.g, die besonderen aber in der Theologie, Rechts-Gelehrsamkeit u. Medicin.' My translation.

16 Seibicke, *Technik*, 127.

17 Ibid.

18 Ibid., 181. Seibicke was able to look at the Kant-index of the academy edition as it was being prepared by Gottfried Martin before completion. Seibicke thus only refers to books I-IX of the academy Edition, due to which his count is low.

19 See Seibicke, *Technik*, 181. Seibicke points out that the term *Technicism*, employed frequently throughout the *Critique of the Power of Judgment,* was most likely a neologism invented by Kant. Seibicke takes the absence of criticism towards Kant's neologisms to signify that they embodied the linguistic dynamics of Kant's epoch.

20 Seibicke, *Technik.* 191. Seibicke proposes that Kant will ultimately combine both the Aristotelian and the Scholastic notion of technics in the sense that his notion of technics deriving from the Scholastic notion will ultimately provide nothing but a particular, objectivized form of the practical capacity of technics in its relation to art (*Kunst*), which derives from *Können.*

21 CPR A70/B96. Guyer and Woods have translated the German word *Technik* as 'technique', for which I have substituted the more encompassing English neologism 'technics', in line with its use in the work of Simondon and Stiegler.

22 Kant, *Lectures on Logic*, 9:18.

23 Seibicke, *Technik*, 183.

24 Kant, *Lectures on Logic.* See Blomberg Logic, §3; Vienna Logic, 795, 820; Hechsel Logic, 114, 117; Dohna-Wundlacken Logic, 781.

25 CPR A708/B736.

26 CPR A719/B747.

27 CPR A712/B740.
28 CPR A727/B755.
29 CPR A707/B735.
30 CPR A729.
31 CPR A732.
32 CPR A733.
33 CPR A734.
34 CPR A735.
35 See Chapter 1.
36 CPR A724/B752-A725/B753.
37 CPR A786/B814. In the *Etymologisches Wörterbuch der Deutschen Sprache,* Wolfgang Pfeifer traces the German word *Sucht* back to the eighth century, where it initially denoted a pathologically heightened desire, need or greed. Its originary meaning as an equivalent of illness/sickness declined from the seventeeth century onwards and was used, probably under the influence of the etymologically unrelated verb *suchen* (seek), to express an intense desire, obsession for and fixation on something. The adjective *süchtig* is further explicated by Wolfgang Pfeifer as something 'filled with a pathological impulse, greedy for something', before being used from the twentieth century onwards to refer to issues of drug addiction. See '*Sucht*' in Pfeifer, Wolfgang et al. Etymologisches Wörterbuch des Deutschen. (1993) digitalized and edited version by Wolfgang Pfeifer in Digitales Wörterbuch der deutschen Sprache. https://www.dwds.de/d/wb-etymwb, accessed on the 29.05.2023. *Sucht* is non-systematically translated as mania in the Cambridge Edition of the *Anthropology from a Pragmatic Point of View* and as lust in the *Critique of Pure Reason*. For reasons of precision and systematicity, I will translate the German *Sucht* as mania, even though in its meaning in Kant it is closer to the contemporary notion of addiction, which however has only been in use from the twentieth century onwards.
38 Kant, Immanuel. *Anthropology from a Pragmatic Point of View,* trans. Robert B. Louden (Cambridge: Cambridge University Press, 2006), AA 7:251.
39 AA 7:266.
40 AA 7:252.
41 AA 7:270.
42 AA 7:266.
43 AA 7:270.
44 CPR B22.
45 AA 7: 268–70.
46 AA 7: 270.
47 CPR Avii. Guyer and Woods here translate *Vermögen* as capacity, but for terminological precision I propose translating it as faculty.

48 CPR B22.

49 Ibid.

50 Ibid.

51 AA 7:253.

52 AA 7:267.

53 CPR Bvii.

54 CPR A709/B737.

55 Ibid.

56 CPR A710/B738.

57 My argument here follows Seibicke, who reads reason's methodology and systematization as merely one particular objectivation of reason's practical faculty, which in Kant is not restricted to the practical domain of material production alone. See Seibicke, *Technik*, 191–2.

58 In this work of building and systematizing, Kant nevertheless reintroduces a difference between the forms of unity employed by reason. Technical unity, he writes, follows from empirical and therefore outer ends, due to which this form is only ever capable of producing contingent unities of aggregation. Architectonic unity, on the other hand, has its end inside itself, according to which its parts are articulated through affinity rather than mere resemblance. Only architectonic unity is universal and necessary and thus properly worthy of the systematicity of science, relegating its technical counterpart to the empirical world of contingency and chance. While the *Critique of Pure Reason* still upholds this distinction between the architectonic and technics, the later *Critique of the Power of Judgment* will revisit this very distinction, as it explicitly aligns systematicity with technics. See Chapter 3.

59 CJ 5:303–4.

60 Ibid.

61 CPR A709/B737.

62 Ibid.

63 CPR A710/B738.

64 CPR A769.

65 CPR A 795/B823.

66 CPR A709/B737.

67 CPR A762/B790.

68 CPR A727/B755.

69 Rousseau, Jean-Jacques. *Émile*, in *Œuvres complètes*, vol. 4 (Paris: Gallimard, 1959–95), 308–9; Rousseau, *Rêveries du promeneur solitaire*, in *Œuvres complètes*, vol. 1 (Paris: Gallimard, 1959–95), 1059; '"Lettre à Malesherbes," 4 January 1762', in *Œuvres complètes*, vol. 1 (Paris: Gallimard, 1959–95), 1132.

Chapter 7

1 CJ 5:303–4.
2 We will encounter this second notion of art and technics as *Können* and practical ability to act and bring forth according to rules in the following chapter.
3 CPR A70/B96, A833/B861, A708/B736, A835/B863, A847/B875.

Part III

1 AA 9:18.
2 CPR 5:170.
3 CJ 20:220.

Chapter 8

1 Kant, Immanuel. *Groundwork of the Metaphysics of Morals*. In *Practical Philosophy*, trans. Mary J. Gregor (Cambridge: Cambridge University Press, 1996), 4:387 (hereafter GMM). One finds this account of the organization of philosophy again in the later, first introduction to the *Critique of the Power of Judgment*. Santotzki argues that Kant might have found inspiration for this organization not only in Lambert and Wolff, but furthermore in the stoic disciplinary organization that he found in Cicero and Seneca. See Santotzki, *Die Bedeutung antiker Theorien für die Genese und Systematik von Kants Philosophie*, 383.
2 GMM 4:387.
3 GMM 4:390.
4 CPR A534/B562.
5 GMM 4:446.
6 See Klinge, Hendrik. *Die moralische Stufenleiter. Kant über Teufel, Menschen, Engel und Gott* (Berlin: Walter de Gruyter, 2018). Klinge's study lucidly demonstrates that and how Kant's moral law extends to other possible moral beings.
7 CJ 5:15.
8 GMM 4:455.
9 GMM 4:412.
10 GMM 4:452.
11 GMM 4:414.
12 Ibid.
13 Ibid.

14 GMM 4:416.
15 Ibid.
16 CJ 20:200.
17 See also Lehmann, Gerhard. 'Kants Nachlaßwerk und die Kritik der Urteilskraft', in *Beiträge zur Geschichte und Interpretation der Philosophie Kants* (Berlin: de Gruyter, 1969), 349.
18 AA 5:26.
19 CJ 5:172.
20 Kant, Immanuel. *The Metaphysics of Morals,* trans. M. J. Gregor (Cambridge: Cambridge University Press, 1991), AA 6:218.
21 This point is also made by Gerhard Lehman in Lehmann, 'Kants Nachlasswerk und die Kritik der Urteilskraft', 349.
22 Stiegler, *TT3,* 193.
23 Ibid.

Chapter 9

1 CJ 5:220.
2 CJ 5:180.
3 The *Critique of Pure Reason* was first published in 1781, with a revised edition released in 1787. There are several important changes that Kant undertook in the revised edition, which is also known as the B-Edition. Substantial changes can be found in the chapter entitled *On the deduction of the pure concepts of the understanding,* in which Kant aimed to prove that and how 'by means of [the categories] alone an object can be thought' (CPR A97). Despite major differences between the A- and B-Deductions, which have become the object of many scholarly debates ever since, Kant claimed that there were in fact no substantial differences, as the two versions are instead argued to complement each other.
4 CPR A105.
5 CPR A116.
6 This point thoroughly troubles Kant's aim of critically restricting reason's (ab)use of logic as an instrument by means of the canon, expounded in Chapter 2. Looking into the workings of synthesis in more detail has revealed that what appeared to be two independent sources of cognition – intuition and understanding – now turn out to be always already in relation to one another with the forms of intuition always already being prefigured by the understanding. Thus, while at the surface Kant critically denies the understanding and its concepts immediate relation to reason's other in favour of intuition, the understanding and its concepts can now be seen to always already loom behind and be active underneath the forms of intuition.

7 CJ 20:220.
8 CJ 20:221.
9 CJ 20:278.
10 Ibid.
11 CJ 5:204.
12 CJ 5:204.
13 CJ 20:215.
14 CJ 5:246. In Guyer and Matthews's translation *Technik* is always translated as technique. But for the sake of terminological consistency, I will from here on adopt the English neologism technics, which one can also find in the English translations of Bernard Stiegler and Gilbert Simondon.
15 CJ 20:215.

Chapter 10

1 See Santozki, *Die Bedeutung antiker Theorien für die Genese und Systematik von Kants Philosophie*, 374; Seibicke, *Technik*, 187.
2 CPR A114.
3 CPR A125. In a footnote to the B-Edition of the *Critique of Pure Reason*, Kant will redefine both terms dynamically. See CPR B446.
4 CPR A114.
5 Guyer, Paul. 'Reason and Reflective Judgment: Kant on the Significance of Systematicity', *Noûs*, On the Bicentenary of Immanuel Kant's Critique of Judgement 24, no. 1 (March 1990): 21.
6 CPR A128.
7 CJ 5:404.
8 CJ 5:406.
9 In the 'Appendix to the Transcendental Dialectic' in the *Critique of Pure Reason*, Kant introduces a regulative use of the ideas of reason, which is charged by Kant with the necessary task of 'bringing unity into particular cognitions as far as possible and thereby approximating the rule to universality'. CPR 647/B675.
10 CPR A179/B221.
11 Lehmann, 'Die Technik der Natur', 292.
12 Aristotle. *The Metaphysics*, trans. H. Lawson-Tancred (London: Penguin Books, 2004), 981a3–5.
13 Aristotle, *The Nicomachean Ethics*, 1139b23.
14 Ibid., 1094a16.
15 Ibid., 1140a18–20.
16 CJ 5:303.

17 CPR A474/B502.
18 CPR A833/B861.
19 Ibid.
20 Santozki, *Die Bedeutung antiker Theorien für die Genese und Systematik von Kants Philosophie*, 366.
21 CJ 5:420.
22 CJ 5:438.
23 Santozki, *Die Bedeutung antiker Theorien für die Genese und Systematik von Kants Philosophie*, 366.
24 CJ 5:438.
25 Santotzki also points out that a differentiation between the two terms is still possible since the architectonic is, properly speaking, a more 'refined' form of technics, the difference of which is marginal for the *Critique of the Power of Judgment.* See Santozki, *Die Bedeutung antiker Theorien für die Genese und Systematik von Kants Philosophie*, 367.
26 CJ 20:215.
27 CJ 20:200–1.
28 CJ 5:412.
29 CJ 5:407.
30 CJ 5:408.
31 CJ 5:407.
32 CJ 5:379.
33 CJ 20:200–1.
34 CJ 20:213.
35 CJ 5:193.
36 See also Blumenberg, Hans. *Schriften zur Technik* (Berlin: Suhrkamp Verlag, 2015), 83.
37 CJ 20:220.

Chapter 11

1 CJ 5:303.
2 CJ 5:220.
3 See also Seibicke, *Technik,* 186.
4 Lehmann, 'Die Technik der Natur', 291.
5 CJ 5:246.
6 CJ 20:221.
7 CJ 20:204.
8 CJ 20:221.
9 CJ 20:211. See also Seibicke, *Technik,* 185.

10 CJ 20:221.

11 Gerhard Lehmann thus argues that the technical power of judgment is nothing else but technical-practical reason. See Lehmann, *Kant's Nachlasswerk*, 355.

Chapter 12

1 Seibicke, *Technik*, 191.

2 CPR 5:170.

3 See Tonelli, Giorgio. 'La formazione del testo della Kritik der Urteilskraft', *Revue internationale de philosophie 8* (1954): 423–48. See Zammito, John H. *The Genesis of Kant's Critique of Judgment* (Chicago/London: University of Chicago Press, 1992), 1–8.

4 CJ 5:383.

5 AA 7:199.

6 Lehmann, Gerhard. *Beiträge zur Geschichte und Interpretation der Philosophie Kants* (Berlin: de Gruyter,1969), 58. Lehmann, Gerhard. 'Neue Perspektiven der Kantforschung', in *Kants Tugenden: Neue Beiträge zur Geschichte und Interpretation der Philosophie Kants* (Berlin: de Gruyter, 1980), 2. 'Kant ist Problemdenker.'

7 CJ 5:170.

Part IV

1 According to Reinhold Jachmann, Kant described the contents of the manuscript known to us as the *Opus Postumum* as both his most important work and as the keystone to his entire doctrinal system. 'Bedauern, ewig bedauern wird aber mit mir jeder Freund der Philosophie, daß den großen Denker Kant das Alter übereilte und ihn an der gänzlichen Vollendung seines philosophischen Systems verhinderte. Mit einer wahren Begeisterung pflegte der unsterbliche Mann oft mit mir über sein letztes Werk zu sprechen, welches nach seiner Aeußerung der Schlußstein seines ganzen Lehrgebäudes seyn und die Haltbarkeit und reelle Anwendbarkeit seiner Philosophie vollgültig documentieren sollte, das aber ganz unvollendet geblieben ist. Es sollte den Uebergang der Metaphysik zur eigentlichen Physik darstellen und auch diesen Titel führen.' See Jachmann, Reinhold Bernhard. *Immanuel Kant geschildert in Briefen an einen Freund* (Königsberg: Nicolovius, 1804), 17–18.

Chapter 13

1 The titel Opus Posthumum first appeared in a letter from Wasianski to Albrecht Krause on 26 June 1883, before Hans Vaihinger referred to the manuscript left behind by Kant under the same name in 1884. See Krause, Albrecht. *Immanuel Kant wider Kuno Fischer zum ersten Male mit Hülfe des verloren gewesenen Kantischen Hauptwerks 'Vom Übergang der Metaphysik zur Physik' vertheidigt von Albrecht Krause. Eine Ergänzung der Populären Darstellung der Kritik der Reinen Vernunft in der Lehre vom Gegenstand und Ding an sich* (Lahr: Schauenburg, 1884), 24. Vaihinger, Hans. 'Zu Kants Widerlegung des Idealismus', in *Straßburger Abhandlungen zur Philosophie* (Freiburg: Mohr, 1884), 150. Concerning the difficulties with the title Opus Postumum, see Brandt, Reinhard. 'Kants Vorarbeiten zum "Übergang von der Metaphysik der Natur zur Physik." Probleme der Edition', in *Übergang: Untersuchungen zum Spätwerk Immanuel Kants* (Frankfurt a.M.: Vittorio Klostermann, 1991), 5.

2 See Karl, Jacqueline. 'Immanuel Kant – der Autor der "mit der Feder in der Hand denkt" Die Arbeitsweise Kants als ein Kriterium für die Neuedition des Opus Postumum', in Annette Sell (Hrsg.), *Editionen – Wandel und Wirkung* (Tübingen: Niemeyer, 2007), 127.

3 Jachmann, Reinhold Bernhard. *Immanuel Kant geschildert in Briefen an einen Freund* in *Immanuel Kant: Sein Leben in Darstellungen von Zeitgenossen* (Darmstadt: Wissenschaftliche Buchgesellschaft, 2012), 17. '. . . Schlußstein seines ganzen Lehrgebäudes.' My translation.

4 OP 21:373. 'Ubergang von den metaph. Anf. Gr. Der Naturwissenschaft zur Physik.'

5 Op 21:465 'Überschritt von der Metaphysik der Körperlichen Natur zur Physik'. My translation. See also Howard, Stephen. *Kant's Late Philosophy of Nature* (Cambridge: Cambridge University Press, 2023), Ch.2.3.

6 For a full account of the reception history of the *Opus Postumum*, see Adickes, Erich. *Kants Opus Postumum* (Berlin: Reuther&Reichard, 1920), 1–35; Förster, Eckhart. 'Introduction to the *Opus Postumum*', in Kant, Immanuel. *Opus Postumum*, trans. Eckhart Förster, Michael Rosen (Cambridge: Cambridge University Press, 1993); Basile, Giovanni Pietro. *Kants 'Opus Postumum' und seine Rezeption* (Berlin: de Gruyter, 2013), 459–98.

7 Lehmann edited both volumes of the *Opus Postumum* as well as Kant's *Lectures* for the academy edition. He is also the editor of the Reclam-edition of the *Critique of the Power of Judgment.*

8 Förster 'Introduction to the *Opus Postumum*' in the Cambridge University Press edition of the *Opus Postumum.*

9 See Tilitzki, Christian. *Die Deutsche Universitätsphilosophie in der Weimarer Republik und im Dritten Reich* (Berlin: Akademie Verlag, 2002), 707. In the

introduction to *Philosophie im Nationalsozialismus*, Hans Jorg Sandkuhler calls Lehmann an aspiring NS-ideologue in the same sentence with Martin Heidegger and Erich Rothacker. See Sandkuhler, Hans Jorg. *Philosophie im Nationalsozialismus* (Hamburg: Felix Meiner Verlag, 2009), 18. Apart from Lehmann's 1940 *Der Einfluß des Judentums auf das französische Denken der Gegenwart*, an explicitly anti-Semitic philosophical text, his 1943 *Die Deutsche Philosophie der Gegenwart* projects an image of contemporary philosophy as 'Aktivistisch, realistisch, existenzialistisch – eine Interpretation jenes Daseins im Kriege, das keine Sicherungen kennt als die Kraft der Gemeinschaft und den Willen zum Siege'. See Laugstien, Thomas. *Philosophieverhältnisse im deutschen Faschismus* (Hamburg: Argument, 1990), 185. After 1945, Lehmann's *Der Einfluss des Judentums auf das französische Denken der Gegenwart* and *Die deutsche Philosophie der Gegenwart* were set on the list of censored books in the Soviet Occupation zone, due to their explicit National Socialist content. It would, however, only take until the 1950s for Lehmann to be commissioned by the now restructured and perhaps too hastily de-Nazified *Deutsche Akademie der Wissenschaften* to edit the fourth section of the Kant edition of Kant's lectures. Lehmann thus found himself politically rehabilitated a mere decade after the end of the war, while two of his books were still censored.

10 Since 2002, the Berlin-Brandenburgische Akademie der Wissenschaften is working on a new edition of the *Opus Postumum*, headed by Eckart Förster, with Jacqueline Karl providing the transcription.

11 In this respect, it is important to distinguish between, on the one hand, the publication details of the *Opus Postumum*, and, on the other, the writing on the *Opus Postumum* undertaken during that same period, a distinction that finds itself doubled in the difference between Lehmann, the Kant-editor, and Lehmann, the Kant-commentator. Lehmann's philosophical contributions remained inseparably entwined with Kantian philosophy, and in particular with the *Opus Postumum*. His most extensive original work on Kant is without doubt his *Habilitationsschrift* from the year 1939, entitled *Kants Nachlasswerk und die Kritik der Urteilskraft*. See Lehmann, Gerhard. 'Kants Nachlasswerk und die Kritik der Urteilskraft', in *Beiträge zur Geschichte und Interpretation der Philosophie Kants* (Berlin: De Gruyter, 1969), 295–373. It was Lehmann's second attempt at acquiring his habilitation. Following a first attempt in 1934, in which Lehmann tried to put forward a sociological analysis of present-day politics under the name *Masse und Macht* (*Mass and Power*), Lehmann's second habilitation attempt appears at first glance as the less explicitly political of the two pieces of writing. At the same time, the first publication following his habilitation was *Die deutsche Philosophie der Gegenwart* in 1943, which attempts a philosophical-theoretical foundation of National Socialism, and is considered as the only 'größeres Berliner Erzeugnis, das eine politische Philosophie des Nationalsozialismus im Rahmen

einer interessanten Philosophiegeschichte entwarf'. See Mehring, Reinhard. 'Tradition und Revolution in der Berliner Universitätsphilosophie', in Von Bruch, Rüdiger. *Die Berliner Universität in der NS-Zeit*. Band II: Fachbereiche und Fakultäten (Stuttgart: Franz Steiner Verlag, 2005), 203. It is thus difficult read Lehmann the Kant-researcher as distinct from Lehmann the National Socialist thinker, with his most comprehensive piece of writing on Kant falling within the same period in which he tried to transform himself into a National Socialist thinker.

12 Lehmann, *Die Technik der Natur*, 212–14.

13 Lehmann, *Kants Nachlasswerk und die Kritik der Urteilskraft*, 295–373.

14 Lehmann, *Die Technik der Natur*, 289.

15 Ibid., 293. Lehmann employs the term 'technische Regionalität' to designate his outline of technics as a whole, which he then distinguishes into four 'Bereiche' (domains, areas, sections). I understand Lehmann's account of domains to mean that these different occurrences of the technical in Kant's thought are ultimately connected, in the sense that they form a landscape, a technical region, within the system of transcendental philosophy, even though how exactly is left open.

16 Lehmann, *Die Technik der Natur*, 293.

17 One can find other speculative proposals in Lehmann's writing concerning the Kantian notion of technics, as for instance his early proposal to investigate Kant in relation to cybernetics, a project undertaken more than half a century later in Yuk Hui's *Recursivity and Contingency*. See Lehmann, 'Neue Perspektiven der Kantforschung', in *Kants Tugenden*, 3; Hui, Yuk. *Recursivity and Contingency* (Washington: Rowman&Littlefield Publishers, 2019).

18 Kant's *Opus Postumum: Online Edition* der Berlin-Brandenburgischen Akademie der Wissenschaften, accessed 01.06.2023, http://telota.bbaw.de/kant_op/.

Chapter 14

1 CJ 5:170.

2 OP 22:545. 'Hieraus läßt sich auch der absolute Wert der Mathematik in Vergleichung mit der Philosophie in Ansehung des Practischen beurtheilen. Der erstere ist der der technisch//practischen (Geschicklichkeit zu Erfindung der Mittel zu beliebigen Zwecken) der andere der moralisch//practischen Vernunft und ist auf den Endzweck der schlechthin (categorisch) gebietend ist, nämlich in Gesinnungen gebesserte Menschen zu schaffen, gerichtet' All translations will in the following be taken from Rosen and Förster's partial translation of the Cambridge edition of the *Opus Postumum* unless stated otherwise.

3 OP 22:545.

4 OP 21:108. 'ein bloßes Kunstproduct des Rechnens und indirect und nur mittelbar (bedingt) auf Zwecke gerichtet dagegen Philosophie auf absolute Zwecke gerichtet ist [. . .] und sich nicht mit Wissenschaft (der Mittel zu Zwecken begnügt).' My translation.

5 OP 22:399. 'Die bewegenden Kräfte der Materie sind also entweder technisch oder blos mechanisch bewegend. – Die letztere sind 1) physisch/mechanisch und setzen die dynamische Möglichkeit als Maschinen voraus.' My translation.

6 Compare Förster, Eckhart. 'Kant's Selbstsetzungslehre', in *Kant's Transcendental Deductions. The Three 'Critiques' and the 'Opus postumum'*, ed. Eckhart Förster (Stanford: Stanford University Press, 1989), 224.

7 OP 22:554. 'Ether ist wirklich weil der Begriff von ihm (mit den Attributen die wir ihm beylegen) die Gesammtheit der Erfahrung möglich macht nicht als Hypothese für wahrgenommene Objecte um ihre Phänomene zu erklären sondern unmittelbar um die Möglichkeit der Erfahrung selbst zu begründen ist er durch die Vernunft gegeben.'

8 OP 22:399. 'die bewegenden Kräfte der Materie sind also entweder technisch oder blos mechanisch bewegend.' My translation.

9 OP 22:48. 'die Thatigkeit desselben ist nach der Analogie der technisch practischen Vernunft.'

10 OP 22:64. 'die Erkenntnis der Mittel zu allen beliebigen Zwecken.' My translation.

11 OP 22:122. 'eine Beziehung seines [des Menschen] Willens auf Zwecke welche in Ansehung seiner bedingterweise nöthigend (neceßitantia) sind: wenn er dieses oder jenes zu bewirken die Absicht hat so muß er dieses oder jenes Verfahren ausüben.'

12 See also Lehman, Kants Nachlasswerk, 352.

13 OP 21:19.

14 OP 21:22.

15 OP 21:11, 12, 13, 14, 16, 19, 21, 22, 23, 24, 26, 31, 32, 43, 44, 45, 47, 51, 53, 54, 78, 95.

16 OP 21:55, 58, 67, 73, 79, 89, 91, 92, 95, 97, 100, 108, 108, 113, 115, 118, 122, 124, 126, 152.

17 OP 21:69, 77, 96.

18 OP 21:19, 22.

19 OP 21:93.

20 OP 21:130.

21 OP 21:373. 'Uebergang von den metaph. Anf. Gr. der Naturwissenschaft zur Physik.'

22 OP 21:27. 'System der Transcendental//Philosophie in drei Abschnitten.'

23 OP 21:54. 'Der höchste Standpunct der Transcendental Philosophie im System der Ideen. Gott die Welt und der seiner Pflicht angemessene Mensch in der Welt.' My translation.

24 OP 21:44.

'Titelblatt u Vorrede

Die Welt als Universum
In allen diesen Objecten ein Maximum Idee ergo unicum in allen 3 Fällen
1. Die theoretisch//speculative
2. Die technisch practische
3. Die moralisch practische Vernunft
Aus Anschauungen, Begriffen a priori u Ideen
Die Idee der Freiheit führt durch den categor. Imperativ auf Gott
1. die speculative
2. die practische
3. die technisch//practische
4. die moralisch//practische Vernunft in einem System.'

25 See CJ 05: 195. In the *Critique of the Power of Judgment* Kant employs the German term 'die Brücke' (bridge) only once, supplemented with 'Kluft' (gulf) and 'Übergang' (transition). See 05: 175–6, 179, 196.

26 OP 21:373. 'Uebergang von den metaph. Anf. Gr. der Naturwissenschaft zur Physik.'

27 OP 21:17. '1) Ubergang von den metaphysischen Anf. Gr der N.W. zur Physik. 2) Ubergang von der Physik zur Tranc: Philos. 3) Ubergang von der Transc. Phil. zum System zwischen Natur und Freyheit. 4) Beschlus von der allgemeinen Verknüpfung der lebendigen Krafte aller Dinge im Gegenverhältnis Gott und die Welt.'

28 OP 21:16. 'Das System der Erkenntnis welches von der Erfahrung (also a priori) dem Formalen nach vorher geht und die Bedingungen der Möglichkeit der Erfahrung überhaupt enthält theilt sich in die zwei Hauptstämme Natur und Freyheit, deren beyde theoretisch und practisch behandelt werden müssen und das Product aus technisch//practischer oder moralisch//practischer Vernunft und ihren Principien (Neigung/Instinct und Sitten/Verstand) hervorgeht.'

29 CJ 5:383.

30 Beiser, Frederick C. *German Idealism: The Struggle against Subjectivism, 1781–1801* (Cambridge, MA: Harvard University Press, 2008), 201.

31 See Förster, Eckhart. *Kant's Final Synthesis* (Cambridge, MA: Harvard Unversity Press, 2000); Emundts, Dina. *Kants Übergangskonzeption im Opus Postumum: zur Rolle des Nachlasswerkes für die Grundlegung der empirischen Physik* (Berlin: de Gruyter, 2004); Hall, Bryan. *The Post-Critical Kant: Understanding the critical philosophy through the Opus Postumum* (New York: Routledge, Taylor Francis Group, 2015).

32 Adickes and Vaihinger both read the *Opus* as internally differentiated and thus as containing the writing to two projects. The first half of the manuscript according to Adickes concerned itself with natural-scientific themes belonging to philosophy of nature, while the second half of the manuscript was considered as Kant's re-engagement with metaphysical and epistemological problems. According to Vaihinger, the first part of the manuscript contains Kant's elaborations on

double-affection, while the second part of the manuscript contains Kant's further developments of transcendental philosophy towards fictionality. See Adickes. *Kants Opus Postumum, dargestellt und beurteilt*; Vaihinger, *Zu Kants Wiederlegung des Idealismus*; Vaihinger, Hans. *Die Philosophie des Als-Ob. System der theoretischen, praktischen und religiösen Fiktionen der Menschheit auf Grund eines idealistischen Positivismus. Mit einem Anhand über Kant und Nietzsche* (Saarbrücken: Dr Müller, 2007).

33 Adickes, *Kants Opus Postumum, dargestellt und beurteilt*, 719.

34 Mathieu, Vittorio. *Kants Opus Postumum* (Frankfurt a. M.:Klostermann, 1989), 250.

35 OP 22:48.

36 Adickes also points out that Beylage V was written on a letter from Wasianski, indicating that it chronologically belongs to Fascicle I rather than V. See Adickes, *Kants Opus Postumum, dargestellt und beurteilt*, 718.

37 The Cambridge University Press edition of the *Opus Postumum* has placed this 'Pseudo-Beylage' back into the contents of Fascicle I. See Kant, *Opus Postumum*. (Cambridge: Cambridge University Press), 210.

38 Mathieu, *Kants Opus Postumum*, 251.

39 OP 21:101.

40 Mathieu, *Kants Opus Postumum*, 251.

Chapter 15

1 While Kant himself speaks of two acts of self-positing in most instances, some readers have found five, or even six acts or steps of self-positing in total. Eckhart Förster argues that self-positing consists of five consecutive steps, which he finds in Fascicle VII before making his way back into the chronologically preceding Fascicles X and XI, to include the topics of self-affection, the hypostatization of space, the problem of the organic body and the exercising of bodily motions in order to solve the difficult problem of self-positing. While he moves backwards into the preceding drafts, he does not move forward to the consecutive Fascicle I. See Förster, *Kant's Final Synthesis: An essay on the Opus Postumum*, 75–116. However, Förster was not the first reader of the *Opus Postumum* who expanded the scope of self-positing. Adickes already argued to have found four, or in fact, six types of self-positing within Fascicles VII and I. Differently to Förster, who finds his five steps of self-positing in the *Opus Postumum* itself and presents us with a fivefold schema of which there is no explicit mention by Kant himself, Adickes argues that his four or rather sixfold schema 'continues' Kant's work. Kant, such is Adickes claim, has not managed to keep these different types of self-positing

apart. Unclarities in the topic itself produced terminological shortcomings, which Adickes aimed to overcome in his contribution. See Adickes, *Kants Opus Postumum, dargestellt und beurteilt,* 628.

2 Adickes, *Kants Opus Postumum*, 668.

3 See Fichte, Immanuel Hermann. *Johann Gottlieb Fichte's Leben und literarischer Briefwechsel* (Leipzig: Brockhaus, 1862), 143–65.

4 Kant 12:371, Briefwechsel Band III (1795–1803). 'ein gänzlich unhaltbares System.' My translation.

5 Förster, *Kant's Final Synthesis,* 77.

6 AA 2:73.

7 AA 2:75.

8 Adickes, *Kants Opus Postumum*, 605.

9 Beck, Jacob S. *Erläuternder Auszug aus den Kritischen Schriften des Herrn Prof Kant* (Riga: Hartknoch, 1794).

10 See also Adickes, *Kants Opus Postumum*, 609.

11 OP 22:353. 'Daß wir nichts einsehen, als was wir selbst machen können. Wir müssen uns aber selbst vorhermachen. Becks ursprüngliches Vorstellen.'

12 Förster, *Kant's Selbstsetzungslehre*, 218.

13 AA 18:615, Reflection 6313.

14 CPR B429.

15 Förster, *Kant's Selbstsetzungslehre,* 218.

16 OP 22:353.

17 CPR B158.

18 CPR Bxviii.

19 Vittorio Mathieu referred to the intensification from the earlier 'combining' and 'having' to 'making' in the *Opus Postumum,* suggesting that Kant was no longer 'satisfied' with a mere agreement of the object of experience with the transcendental structure of the subject. Mathieu then goes on to propose that the 'we' in Kant's formulation '[. . .]we must make ourselves'. (OP 22:353) does not refer to the faculty of understanding in general. It rather refers to the individual scientist and researcher, proposing that the experience of the specific nature which we encounter, has to be made by each human in general and by the scientist in specific. This leads Mathieu to the proposition that Kant's explicit intention of providing a 'bridge' from metaphysics to physics in fact coincides with the amplification of the productive power of imagination of the individual scientist rather than with the faculty of the understanding. See Mathieu, Vittorio. 'Erfinderische Vernunft in Kants Opus Postumum', in *Übergang: Untersuchungen zum Spätwerk Immanuel Kants* (Frankfurt a. M.: Vittorio Klostermann, 1991), 65–76.

20 See Lehmann, *Kants Nachlaßwerk und die Kritik der Urteilskraft*, 358.

21 OP 22:70.
22 OP 22:73. 'die Bestimmung seiner selbst als Function seiner selbst als *vis representativa*.' My translation.
23 OP 22:73.
24 OP 22:73, 43, 58, 66, 67, 68, 77, 79, 83, 85, 86, 88, 90, 93, 95, 97, 98, 102, 115.
25 OP 22:43. 'die Vorstellung seiner selbst (apperception) wodurch das Subject sich selbst zum Object macht.'
26 OP 22:115. 'entspringt nicht aus einem vorhergehenden Akt.'
27 OP 22:73. 'geht von innen hinaus mit etwas was sie selbst setzt.' My translation.
28 OP 22:95.
29 OP 22:93. 'das Formale des Urteilens nach der Regel der Identität.' My translation.
30 OP 22:67.
31 OP 22:79.
32 CPR B157.
33 OP 22:72. 'Nur dadurch dass das Subject sich selbst Object wird *schreitet* die Apperzeption zur Apprehension und *schreitet* von der Metaphysik zur Transzendentalen Philosophie, vom Analytischen zum Synthetischen a priori.' My translation and italicization.
34 OP 22:72.
35 OP 22:91.
36 OP 22:96.
37 OP 22:82.
38 OP 22:96. 'das logische Bewußtsein führt zum Realen.'
39 OP 22:82.
40 OP 22:91. 'Die Prinzipien der Möglichkeit Erfahrung anzustellen liegen identisch im Vorstellungsvermögen überhaupt (facultas repraesentativa) woraus die Formen der Urteilskraft analytisch und die Prinzipien der Vernunft apodiktisch hervorgehen.' My translation. The editors Lehmann and Buchenau mention that 'analytically' replaced the initial 'synthetically'.
41 Ibid.
42 OP 21:98. 'Noch vor einer Selbstbestimmung im Raume und der Zeit (obzwar, zum Behuf derselben).' My translation.
43 CPR A62/B87.
44 OP 21:43.
45 OP 21:93. See also 21:84: 'Transcendental philosophy is the self-creation (autocracy) of ideas, into a complete system of the objects of pure reason.'
46 CPR A409/B435.
47 CPR A409/B436.
48 OP 22:15. 'Geht voran mit der Entwerfung ihrer Formen.' My translation.
49 OP 21:88. 'Ideen gehen vor den Erscheinungen in Raum u. Zeit vorher.'

50 OP 21:85–6. 'Gedankenwesen, subjective Producte der eigenen Menschenvernunft, die das Subject auf sich selbst bezieht.' My translation.
51 OP 21:93. 'Ideen sind Gesetze des Denkens die das Subject ihm selbst vorschreibt. Autonomie.'
52 OP 21:79. 'Ideen sind nicht Begriffe sondern reine Anschauungen nicht discursive sondern intuitive Vorstellungen, denn es ist nur Ein solcher Gegenstand.'
53 OP 21:51. 'Ideen sind a priori durch reine Vernunft geschaffene Bilder (Anschauungen) welche vor der Erkenntnis der Dinge vorher blos subjective Gedankendinge und die Elemente der letzteren vorhergehen. Sie sind die Urbilder (prototypa) [. . .].'
54 CPR A313/B370.
55 CPR A568/B596.
56 CPR A569/B597.
57 CJ 5:232.
58 OP 21:51.
59 OP 21:40. 'ein jeder dieser Gegenstände ist schlechthin Einer (unicum).'
60 OP 21:11. 'Both are a maximum: the one determined according to degree (qualitative), the other according to volume [or] space (quantitative); the one as object of pure reason, the other as sense object. Both are infinite: the first as magnitude of appearance in space and time; the second according to degree (*virtualiter*), as limitless activity with regard to forces (mathematical or dynamic magnitudes of sense objects). One as *thing in itself* or *appearance*.'
61 OP 21:11. 'unendlich; der erste als Größe der Erscheinung im Raum und der Zeit der zweite dem Grad nach (virtualiter).' My translation.
62 Förster, *Kant's Final Synthesis*, 162.
63 OP 22:554. 'It is given by reason, not as a hypothesis for perceived bodies, for the purpose of explaining their phenomena, but rather, immediately, in order to found the possibility of experience itself.' Kant then goes on to explain the unique and singular proof of the existence of ether.
64 OP 21:93. 'hypothetisch (problematisch oder assertorisch) sondern apodictisch indem sie [reason/*Vernunft*] sich selbst schafft.' My translation.
65 OP 21:43. 'ebenso können wir sagen sie [alle Gegenstände] müssen ihrer Realität nach in der Welt angetroffen werden.' My translation.
66 OP 22:43. 'in der Verschiedenheit des Verhältnisses wie das den Sinnengegenstand apprehendierende Subject zur Bewirkung der Vorstellung in ihm afficirt wird.'
67 OP 21:37. 'Gott u die Welt sind beide Objecte der Transc. Phil und (Subject, Praed u copula) ist der denkende Mensch. Das Subject der sie in einem Satze verbindet. – Dieses sind logische Verhältnisse in einem Satze nicht die Existenz der Objecte betreffend sondern blos das Formale der Verhältnisse diese Objecte

zur Synthetischen Einheit zu bringen Gott, die Welt, und Ich der Mensch ein Weltwesen selbst, beide verbindend.'

68 OP 21:29. 'Gott, die Welt, und der Mensch als Person d.i. als Wesen das diese Begriffe vereinigt.'

69 OP 21:34. 'Gott, die Welt und was beide zu einem System vereinigt das denkende einwohnende Prinzip des Menschen (mens) in der Welt.'

70 CPR A327/B384.

71 CPR A334/B391.

72 Kant, *Jäsche Logik*. AA 9:25. Repeated in a letter of Kant to Carl Friedrich Stäudlin (4 May 1773), AA 11:574.

73 Kant 9:25.

74 Bermes, Christian. *Welt: Vom metaphysischen natürlichen Weltbegriff* (Hamburg: Felix Meiner Verlag, 2004), 64.

75 Foucault, Michel. *Introduction to Kant's Anthropology*, transl. Roberto Nigro, K. Briggs (Los Angeles: Semiotext(e) Foreign Agents Series, 2008), 88.

76 Fink, Eugen. *Einleitung in die Philosophie* (Würzburg: Königshausen u Neumann, 1985), 81. '[. . .] ist Kant dennoch der Höhepunkt der philosophischen Frage nach der Welt in der gesamten Tradition der abendländischen Metaphysik: bei ihm vollzieht sich erstmalig die prinzipielle Einsicht und Begründung dafür, daß die Welt *kein Seinendes, kein Ding ist*, daß jeder Versuch sie dinghaft zu denken, ob als endliche oder unendliche Größe, die menschliche Vernunft in einen unauflöslichen Widerspruch verwickelt.' My translation.

77 Foucault, *Introduction to Kant's Anthropology*, 33.

78 See Fink, *Einleitung in die Philosophie*, 81.

79 Bermes, *Welt*, 56.

80 *Announcement of the Organisation of his Lectures in the Winter Semester* 1765–1766, translated in Caygill, *A Kant Dictionary*, 214.

81 The encompassing introduction by Rink contains the first published reference to Kant's work on the manuscript of his transition project, when Rink writes 'gelänge es doch dem ehrwürdigen Urheber dieser phys. Geographie noch seinen Übergang von der Metaphysik der Natur zur Physik bekannt zu machen! Auch über diesen Gegenstand würde man dort, wie ich weiß, manche scharfsinnige Bemerkung vorfinden.' AA 9:221.

82 AA 9:157.

83 AA 9:158.

84 AA 9:158.

85 AA 9:158.

86 AA 9:157.

87 AA 2:443.

88 This point on the entwined relation between the two disciplines is also repeated by Kant in *Physical Geography*, AA 9:157. The inner relation of anthropology

to geography must have escaped Heidegger, who mainly references Kant's *Anthropology*, but did not seem to have much to say about Geography.

89 Bermes, *Welt*, 61. 'das je schon als spezifische, natürliche Erfahrung erworben ist und in dieser Form auch kultiviert werden kann'. My translation.

90 See Bermes, *Welt*, 65.

91 CPR A838/B866.

92 CPR A838/B866.

93 Kant, *Lectures on Logic*, AA 9:24.

94 Ferrarin, Alfredo. *The Powers of Pure Reason; Kant and the Idea of Cosmic Philosophy* (Chicago: The University of Chicago Press, 2015).

95 CPR A838/B866.

96 Kant, *Lectures on Logic*, AA 9:24.

97 CPR A839/B867.

98 In the *Critique of Pure Reason,* the three questions of speculative and practical philosophy are also specified as: '1. What can I know? 2. What should I do? 3. What may I hope?' CPR A805/B833.

99 CPR A840/B868.

100 Kant, *Lectures on Logic*, AA 9:24.

101 OP 21:117. 'Philosophie (doctrina sapientia) ist nicht eine Kunst von dem was aus dem Mensch zu machen ist sondern was er aus sich selbst machen soll (sapere aude). Versuche dich deiner eigenen Vernunft zu deinem wahren Zwecke zu bedienen.'

102 AA 7:119.

103 OP 21:94. 'Was macht der Mensch aus sich selbst?'

104 OP 21:40. 'Wenn Gott ist so nur einer. Ist eine Welt in metaphysischer Bedeutung so ist nur Eine und ist der Mensch so ist es das Ideal Urbild Prototypon Eines der Pflicht adäquaten Menschen.'

105 OP 21:48. 'ist eine bloße Idee der reinen Vernunft der categorische Imperativ bewährt ihm seine Realität und er ist insofern Noumenon.' My translation.

106 OP 21:48. See also Mathieu, *Kants Opus Postumum,* 264.

107 OP 21:49. 'Ich der Mensch bin ein Weltwesen und Erscheinung im Raum und der Zeit.' My translation.

108 Already in relation to the earlier *Anthropology*, Foucault had pointed out that 'man is neither a *homo natura*, nor a purely free subject; he is caught by the syntheses already operated by his relationship to the world'. Foucault, *Introduction to Kant's Anthropology*, 54–5.

109 OP 21:61.

110 OP 21:27 'Der medius terminus (copula) im Urtheile ist hier das Urtheilende Subject (das denkende Weltwesen, der Mensch, in der Welt.) Subiect, Praedicat, Copula.'

111 OP 21:26. 'Not a system of nature but of thought.'

112 OP 21:90. 'Wissenschaft der Formen ähnlich unter welchen wenn sie gegeben werden sollten sie allein erscheinen müßten.' My translation.

113 CPR A11/B25.
114 CPR A13/B27.
115 OP 21:63–4. 'eine Wissenschaft welche (objectiv) Grundsätze über Objekte sondern über das Subject der Erkenntnis desselben Umfang und Grenzen seines Wissens vorträgt.' My translation.
116 OP 21:75. 'ein Vermögen sich selbst synthetisch in seiner durchgängigen Bestimmung als in einem System der Ideen a priori gegeben vorzustellen.' My translation.
117 OP 21:79. 'Transc. Philos. ist die Autonomie der Ideen in so fern sie unabhängig von allem Empirischen ein unbedingtes Ganze ausmachen und die Vernunft sich selbst zu diesem als einem absonderlichen System constituiert.'
118 OP 21:21. '. . . das Denken wodurch wir uns die Gegenstände selbst durch synthetische Erkenntnis a priori aus Begriffen selbst machen und der gedachten Gegenstände subjective Selbstschöpfer sind.'
119 OP 21:97. 'Transc. Phil. ist das formale Princip sich selbst als Object der Erkenntnis systematisch zu constituieren.' My translation.
120 OP 21:98. 'noch vor seiner Selbstbestimmung im Raume u. der Zeit doch zum Behuf derselben.' My translation.
121 OP 21:7. 'Die Transc: Philosophie geht vor der Behauptung der gedachten Dinge Voraus als ihr Urbild wohin sie gestellt werden müssen.'
122 Förster, *Kant's Final Synthesis*, 163.
123 The source and layout of the concepts of reflection are not disclosed in the *Critique of Pure Reason*. Kant merely refers to logicians '*von ehedem*' (from before), meaning that they have been handed down by tradition. See CPR A266/B322.
124 CPR A260/B316.
125 CPR A262/B318.
126 CPR A268/B324.
127 On the aporia of judgment, see Chapter 2.
128 CPR A261/B317.
129 CPR A268/B324.
130 CPR A268/B324.
131 CPR A269/B325.
132 OP 21:7. 'geht vor der Behauptung der gedachten Dinge Voraus als ihr Urbild wohin sie gestellt werden müssen.' My translation.
133 Phenomena are 'appearances, to the extent as objects they are thought in accordance with the unity of the categories'. CPR A248.
134 OP 21:43. Foucault points out that finally, in the *Opus Postumum*, Kant conceded to thinking the absolute from the starting point of man. Foucault, *Introduction to Kant's Anthropology*, 78.
135 Förster, *Kant's Final Synthesis*, 163.

136 OP 21:21. 'Gott und die Welt sind Ideen der moralisch//practischen und technisch//practischen auf Sinnenvorstellungen gegründeten Vernunft.'

137 OP 21:78. 'Transc: Phil. ist der Act des Bewußtsyns dadurch das Subject seiner selbst Urheber wird und dadurch auch von dem ganzen Gegenstande der technisch//practischen und moralisch//pracischen Vernunft in Einem System in Gott alle Dinge als in Einem System zu ordnen.'

138 OP 22:53. 'Das Subject bestimmt sich selbst 1) durch technisch//practische 2) durch moralisch//practische Vernunft und ist sich selbst ein Gegenstand von beyden Die Welt und Gott. Das erste im Raum u. der Zeit als Erscheinung. Das zweyte nach Vernunftbegriffen d.i. einem Princip des categorischen Imperativs.'

139 OP 21:19. 'Gott und die Welt. Ein System der Transcendentalphilosophie von Technisch theoretischer und moralisch//practischer Vernunft.'

140 OP 22:64. 'Gott u die Welt. Jede ein der Qualität besonderes Ganze. Einzeln und in Verbindung. Technisch//practische und moralisch//practische Vernunft. Jedes ein Ganzes und beide zusammen in Verbindung (das All der Objecte u des Subjects).' My translation.

141 OP 22:399.

142 OP 21:4. 'Der Weltraum wird als eine allgemeine Basis der Körperwelt gedacht.' My translation.

143 OP 21:24. 'Ein Vernunftbegriff der Freyheit in so fern in ihm ein Princip der Verknüpfung des Mannigfaltigen liegt das nur einer Person zukommt.' My translation.

144 OP 21:13. '. . . das Ganze aller Gegenstände der Sinnenvorstellung.'

145 OP 21:52. 'Das eine geht auf Kunst u. Geschicklichkeit (nach beliebigen Zwecken) das andere auf Weisheit auf den Endzweck.'

146 See Höffe, Ottfried. *Kants Kritik der reinen Vernunft: eine Philosophie der Freiheit* (München: Beck, 2012).

147 Lehmann, *Kants Nachlasswerk und die Kritik der Urteilskraft*, 352. 'die technisch-praktische Vernunft eben diesen Weltbezug des theoretischen Naturbezugs [bezeichnet].' My translation.

Chapter 16

1 Lehmann, *Kants Nachlasswerk und die Kritik der Urteilskraft*, 351.

2 For an in-depth discussion of moral-practical self-positing, see Förster, *Kant's Final Synthesis*, 117–47.

3 OP 21:23. 'Das denkende Subject schafft sich [. . .] eine Welt als Gegenstand möglicher Erfahrung im Raum und der Zeit.' My translation.

4 OP 22:12. 'Nicht der Raum als Objekt wird angeschaut sondern er ist die Synthesis des Mannigfaltigen im vorstellenden Subjekt selbst In dieser Vorstellungsart wodurch das Subjekt sich selbst constituirt.'
5 OP 22:88. 'Raum und Zeit in der Anschauung sind nicht Dinge sondern actus der Vorstellungskraft sich selbst zu setzen, wodurch das Subject sich selbst zum Object macht.'
6 OP 22:96. 'das Object der reinen Anschauung vermittelst welcher das Subject sich selbst setzt ist unendlich nämlich Raum und Zeit.'
7 Lehmann, *Kants Nachlasswerk und die Kritik der Urteilskraft*, 359.
8 CPR A15/B29.
9 See also Beiser, *German Idealism: The Struggle against Subjectivism.*
10 CPR A20/B34.
11 CPR A20/B34.
12 CPR A20/B34.
13 CPR A20/B34.
14 OP 22:15. 'Raum und Zeit sind nicht von unseren Vorstellungen unterschiedene Dinge als Gegenstände unserer Anschauung sondern sind Anschauungen selbst; doch nicht empirische (Wahrnehmungen) des Sinnenobjects denn alsdann würden sie Vorstellungen sein welche von einer anderen Kraft die uns afficirte, und als Ursache zukämen sondern sind Producte unserer eigenen Vorstellungskraft.' My translation.
15 OP 22:37.
16 OP 22:37. 'Raum und Zeit sind Producte (aber primitive Producte) unserer eigenen Einbildungskraft mithin selbstgeschaffene Anschauungen.'
17 OP 22:76. 'Die Vorstellung derselben [Raum und Zeit] ist ein Act des Subjects selbst und ein Product der Einbildungskraft für den Sinn des Subjects.'
18 OP 22:42. 'Raum u Zeit sind nicht discursive Vorstellungen durch Begriffe sondern intuitive nicht Gegenstände der Anschauung des Mannigfaltigen sondern Anschauung selbst und zwar nicht des Allgemeinen sondern des Einzelnen (es ist Ein Raum und Eine Zeit) Er ist kein wahrnehmbarer (apprehensibeler) Gegenstand sondern Product des Vorstellungsvermögens als Selbstthätigkeit (Spontaneitas nicht Receptivitas) das aspectabile als cogitabile vorzustellen.' My translation.
19 OP 22:45. 'Raum und Zeit sind keine spührbare (apprehensibele) gegenstände die ausser meiner Vorstellung existieren sindern selbst Geschöpfe meines Vorstellungsvermögens.' My translation.
20 CPR A99.
21 OP 22:443–444. 'Der Verstand fängt nicht vom Object sondern von seinem eigenen Subject an die Sinnenvorstellung ihrer Form nach zu construieren d.i. synthetisch a priori das Mannigfaltige derselben in der Einheit desselben nach einem Prinzip darzustellen welches eine mathematische Operation desselben ist [. . .] Die

unbedingte Eineit des Mannigfaltigen in der Anschauung ist nicht dem Subject von einem anderen Gegenstande gegeben sondern durch dasselbe gedacht. Raum u. Zeit sind nicht als Verstandesbegriffe Anticipationen der Wahrnehmung sondern Formen der Gegenstände in der Erscheinung.'

22 CPR A176/B218.

23 OP 22:99. 'das Formale dieser Anschauung ist Eines und Alles zusammengefaßt ist die Vorstellung von Raum und Zeit welche nicht analytisch aus Begriffen sondern synthetisch durch Construction der Begriffe eine Unendlichkeit (unbegrenzte Größe) vorstellt.'

24 OP 22:69. 'Ist reine nicht empirische Anschauung welche nicht wie in der Logik vom allgemeinen zum besonderen analytisch sondern synthetisch vom All des Inbegriffs des Mannigfaltigen zum Einzelnen synthetisch sich selbst a priori constituirt und blos das Formale des Mannigfaltigen der Anschauung des Subjects in der Totalität d.i. das Unbedingte Ganze der Anschauung enthält.' My translation.

25 OP 22:116. 'Es ist aber ausser der Sinnenvorstellung noch ein Erkenntnisvermögen welches nicht blos Receptivität sondern auch Spontaneität (als oberes Erkenntnisvermögen) enthält nämlich Verstand, Urtheilskraft und Vernunft und diese kann technisch// Anschauung construierende oder moralisch//practische Vernunft seyn beyde a priori.'

26 Lehmann, *Technik der Natur*, 293.

27 In Kant, *Theoretical Philosophy after 1781*, 271–337; AA 8.

28 AA 8:192.

29 AA 8:192.

30 CPR A729.

31 Kant, *Prolegomena to Any Future Metaphysics with Selections from the Critique of Pure Reason*, Revised Edition, trans. and ed. Gary Hatfield (Cambridge: Cambridge University Press, 2004), §7, AA 4:281.

32 CPR A723/B751.

33 CPR A136/B175.

34 See Chapter 2 for the discussion on the methodological difference between mathematics and philosophy.

35 Förster, *Kant's Final Synthesis*, 105. The consecutive fourth and fifth steps are considered by Förster to consist of inserting the concept of ether into the sensible, and, finally, performing bodily actions initiated by the subject within this sensible world of forces. See 111–12.

36 Förster, *Kant's Final Synthesis*, 105, 109.

37 See Howard, Stephen. 'The transition within the transition: the Übergang from the Selbstsetzungslehre to the ether proofs' in *Kant's Opus Postumum* in *Kant-Studien* 110, 4, 2019, 595–617.

38 I will in the following also engage with Fascicles X and XI, but only to lay open the details of Kant's discussion on self-affection. The transition, according to my argument, is located in Kant's discussion of the system of transcendental philosophy in Fascicle I.

39 CPR B67.

40 CPR B67.

41 CPR B68.

42 CPR B68.

43 OP 22:420. 'Our sensible intuition is, initially, not perception (empirical representation with consciousness), for a principle of positing oneself and of becoming conscious of this position precedes it; and the form[s] of this positing of the manifold, as thoroughly combined, are the pure intuitions, which are called space and time (outer and inner intuition) [. . .].'

44 OP 22:368. 'das zusammensetzende Subject erscheint sich selbst in der Zusammensetzung nach Prinzipien a priori.' For the relation between second-order appearances and the possibility of physics see Beiser, *German Idealism: The Struggle against Subjectivism, 1781–1801*, 203–4; Lehmann, *Kants Nachlasswerk und die Kritik der Urteilskraft*, 366–7.

45 CPR B155.

46 OP 22:320.

47 OP 22:367. 'Erscheinung des sich selbst affizierenden Subjects.'

48 OP 22:390. 'Also ist nicht ein empirisches System (denn das enthält einen Wiederspruch) sondern ein System empirischer Vorstellungen was a priori in dem Ubergange zur Physik zum Grunde liegt der Form nach in der Erscheinung des afficirten und sich selbst afficirenden Subjects, nicht sie durch Erfahrung aufzuheben sondern für die Erfahrung zuerst hinein zu tragen.'

49 See Vaihinger, 'Zu Kants Widerlegung des Idealismus', in *Strassberger Abhandlungen zur Philosophie*, 145; Adickes, *Kant's Opus Postumum*, 293; Kemp Smith, Norman. *A Commentary to Kant's Critique of Pure Reason* (Charleston: Forgotten Books, 2012), 607.

50 Lehmann, *Kants Nachlasswerk und die Kritik der Urteilskraft*, 364.

51 CPR A20/B34.

52 This position is also put forth by Lehmann, *Technik der Natur*, 293.

53 OP 22:477. '[. . .] das Subjekt affiziert sich selbst und nimmt das wahr "was es selbst in der (*sic*!) empirischen Anschauung (Wahrnehmung) hineingelegt hat.[the subject affects itself and perceives what it has itself placed into empirical intuition (perception)].'" My translation.

54 OP 22:363. 'Erscheinung von einer Erscheinung ist das wodurch das subjective objective gemacht wird.' My translation.

55 OP 22:48.

56 OP 21:58. 'Im Raum und der Zeit [. . .] zusamt dem Subject (der Mensch) welches selbst ein Theil der Welt ist.' My translation.

57 Lehmann, *Technik der Natur*, 293.

Chapter 17

1 GMM 4:416.

2 CJ 5:172.

3 OP 21:43.

4 OP 21:31 'Ein Cosmotheoros der die Elemente der Welterkenntnis a priori selbst schafft aus welchen er die Weltbeschauung als zugleich Weltbewohner zimmert in der Idee.'

5 OP 21:41.

6 CPR A838/B866.

7 Lehmann, G. 'Die Technik der Natur', in *Beiträge zur Geschichte und Interpretation der Philosophie Kants* (Berlin: Walter de Gruyter & Co, 1969). First published in 1938 in: Forschungen und Fortschritte, 14. Jg. Nr. 18, pp. 212–21. p. 289.

8 Hui, Yuk. *The Question Concerning Technology in China* (Falmouth: Urbanomic, 2016).

9 Hui, Yuk. 'Cosmotechnics as Cosmopolitics', *e-flux* #86 (2017): 1–11.

Epilogue

1 See Howard, Stephen. *Kant's Late Philosophy of Nature* (Cambridge: Cambridge University Press, 2023), Chapter 2.

2 1798 letter to Garve in AA 12:257; 1798 letter to Kiesewetter in AA 12.

3 Kuno Fischer was the first to claim in his 1884 *Das Streber- und Gründertum in der Literatur. Vade Mecum für Herrn Pastor Krause in Hamburg* that there is no need for a bridge or transition, since there is no abyss. See Förster, Eckhart. *Kant's Final Synthesis: An Essay on* the *Opus Postumum* (Cambridge, MA: Harvard University Press, 2002), 48ff for a historical and conceptual discussion of the gap and the reception of this gap within post-Kantian philosophy and Howard's *Kant's Late Philosophy of Nature* on a critique of Förster.

4 See Howard, *Kant's Late Philosophy of Nature*, Ch. 2.4.

5 According to Reinhold Jachmann, Kant described the *Opus Postumum* as both his most important work and as the keystone to his entire doctrinal system. 'Bedauern, ewig bedauern wird aber mit mir jeder Freund der Philosophie, daß den großen Denker Kant das Alter übereilte und ihn an der gänzlichen Vollendung seines

philosophischen Systems verhinderte. Mit einer wahren Begeisterung pflegte der unsterbliche Mann oft mit mir über sein letztes Werk zu sprechen, welches nach seiner Aeußerung der Schlußstein seines ganzen Lehrgebäudes seyn und die Haltbarkeit und reelle Anwendbarkeit seiner Philosophie vollgültig documentieren sollte, das aber ganz unvollendet geblieben ist. Es sollte den Uebergang der Metaphysik zur eigentlichen Physik darstellen und auch diesen Titel führen.' See Jachmann, Reinhold Bernhard. *Immanuel Kant geschildert in Briefen an einen Freund* (Königsberg: Nicolovius, 1804), 17–18.

6 Kant quotes Virgil's Aenead 'Styx interfusa coërcet' 6.436–9 at various points throughout the manuscript. See OP 22:491, 22:487, 22:489, 21:175, 21:139, etc.

7 Hesiod, *Theogony*, trans. M. L. West (Oxford: Oxford Worlds Classics, 2009), 361; 385.

8 Hesiod, *Theogony*, 383–403; Aristotle, *Metaphysics*, 983b29–984a4.

9 See also Agamben, Giorgio. *The Sacrament of Language: An Archeology of the Oath*, trans. Adam Kotsko (Stanford: Stanford University Press, 2011), 19.

10 The punishment for the Gods of breaking an oath was severe, and included, among others, the loss of their voice for nine years.

11 See Weinrich, Harald. *Lethe, Kunst und Kritik des Vergessens* (München: C.H. Beck, 1997), 18.

12 *Lethe* (λήθω), ancient Greek for forgetting, oblivion and concealment. The concept of alatheia (ἀλήθεια), the ancient Greek notion of truth is derived from the same word root, due to which Heidegger famously interpreted it as unconcealment. See Weinrich, *Lethe, Kunst und Kritik des Vergessens.*

13 Matthias Friedrich Nicolovius (1768–1836) owned a bookshop and publishing house in Königsberg that published many of Kant's writings. Kant, who had only a small collection of books himself, had an agreement with the bookshop to provide him any and all books he might need for his studies. See Rink, Friedrich Th. *Fragmente aus Kant's Leben: ein biographischer Versuch* (Königsberg: Hering & Haberland, 1807), 54. After Kant's death, the biographies by Borowski, Jachmann and Wasianski were published by Nicolovius, due to which they are often referred to as the Nicolovius-biographers.

14 See Jachmann, *Immanuel Kant geschildert in Briefen*, 178. 'Schon vor acht Jahren fand ich ihn etwas verändert, obgleich er sich an einzelnen Tagen, wenn die Funktionen der Natur gut vonstatten gingen, noch ganz in seiner vormaligen Geisteskraft zeigte. Seit dieser Zeit aber ward die Abnahme seiner Kräfte merklicher. Vor vier Jahren fing er schon an, sich eines Gedankenzettels zu bedienen [. . .]. Auf diese Blättchen schrieb er endlich jede Kleinigkeit auf, die ihm von andern gesagt oder ihm selbst eingefallen war. [. . .] So schwand allmählich die Kraft des größten Denkers bis zur völligen Geistesohnmacht hin.'

15 Wasianski, E. A Ch. 'Immanuel Kant in seinen letzten Lebensjahren', in Borowski, Jachmann und Wasianski. *Immanuel Kant. Sein Leben in Darstellungen von*

Zeitgenossen (Darmstadt: Wissenschaftliche Buchgesellschaft, 2012), 206. 'Es schien, als ob das, was *Kants* ganzes Leben hindurch ein Fehler an ihm, obgleich im unmerklichen Grade gewesen, nämlich, eine besondere Art von Vergeßsamkeit in Dingen des gemeinen Lebens, nun mit den Jahren einen höheren Grad erreicht hat. Er selbst gestand, daß er sich diesen Fehler sehr oft habe zuschulden kommen lassen, und führte als Beleg aus den frühsten Jahren seines Lebens folgende Geschichte an. Als ein ganz kleiner Knabe hielt er sich, wie er aus der Schule kam, gewisser, leicht zu erratender Ursachen wegen, einige Augenblicke unter einem Fenster auf, hing seine Bücher an den Ladenriegel und vergaß sie wieder abzunehmen. [. . .] Noch in den spätern Jahren seines Lebens [. . .] machte [er] auch kein Geheimnis daraus, daß er sonst schon vergeßsam gewesen sei. Was früher sich seltener ereignete, traf nun im Alter häufiger ein.'

16 Borowski, Ludwig E. 'Leben und Charakter Immanuel Kants', in Borowski, Jachmann und Wasianski. *Immanuel Kant. Sein Leben in Darstellungen von Zeitgenossen* (Darmstadt: Wissenschaftliche Buchgesellschaft, 2012), 65. 'In seiner frühen Kindheit sei er, das gestand er selbst zu, in manchen Dingen, die besonders den Schulfleiß nicht affizierten, vergeßsam gewesen; – in ganz unbeträchtlichen und von ihm für unbeträchtlich gehaltenen Dingen war er es wohl auch in höhern Jahren; – zuletzt machte er sich Denk- und Erinnerungszettel'.

17 See also Basile, Giovanni Pietro. *Kants 'Opus Postumum' und seine Rezeption* (Berlin: de Gruyter, 2013), 457.

18 See Hasse, Johann G. *Merkwürdige Aeusserungen Kant's von einem seiner Tischgenossen* (Königsberg: Hering, 1804), 14.

19 Basile, *Kants 'Opus Postumum' und seine Rezeption*, 458. See Basile, *Kants 'Opus Postumum' und seine Rezeption*, 459–81, for an account of how the manuscript was found again until its eventual publication in 1936/1938.

20 Fischer, Kuno. *Geschichte der neuern Philosophie. Bd. 3: Entstehung und Begründung der kritischen Philosophie. Die Kritik der reinen Vernunft. Immanuel Kant. Entwicklungsgeschichte und System der kritischen Philosophie* (Mannheim: Bassermann, 1860), 83. 'Hinfälliger Zustand.' My translation.

21 Fischer, Kuno. 'Das Streber- und Gründerthum in der Literatur. Vademecum für Hrn. Pastor Krause in Hamburg', in *Beilage zur Allgemeinen Zeitung (München)* Nr. 148 from 25.05.1884, 2171. 'Absterbenden Geist.' My translation.

22 Schepelmannn, Maja. *Der Senile Kant? Zur Wiederlegung einer populären These* (Paderborn: Brill, 2018), 43.

23 Vaihinger, Hans. 'Eine Blattversetzung in Kants Prolegomena'. *Philosophische Monatshefte 15* (1879), 321–32, 513–32.

24 Schopenhauer, Arthur. *Die Welt als Wille und Vorstellung, Bd 1: Vier Bücher, nebst einem Anhang, der die Kritik der Kantischen Philosophie enthält* (Leipzig: Brockhaus,

1819), Buch 4, §62. 'Nur aus Kants Altersschwäche ist mir seine ganze Rechtslehre [. . .] erklärlich.'

25 Letter of Schopenhauer to Rosenkranz on the 24.08.1837. See Kant's *Critique of Pure Reason*, ed. Karl Rosenkranz (Leipzig: Voss, 1838), XI–XIV. '[. . .] Menschenfurcht, entstanden durch Altersschwäche, welche nicht nur den Kopf angreift, sondern bisweilen dem Herzen jene Festigkeit nimmt [. . .].'

26 Ludwig, Bernd. *Kants Rechtslehre* (Hamburg: Meiner, 1988), 4.

27 Brandt, Reinhard. *Kritischer Kommentar zu Kants Anthropologie in pragmatischer Hinsicht* (1798) (Hamburg: Meiner, 1999), 8.

28 Kühn, Manfred. *Kant. Eine Biographie* (München: Beck, 2003), 467.

29 Adickes, *Kants Opus Postumum,* 593. Further references to Kant's senility on 84, 148–53, 593, 719–23.

30 Ibid., 719.

31 Ibid. '[. . .] ohne Zweifel weil es Kant an der erforderlichen Fähigkeit fehlt, ein größeres Ganzes nicht etwa nur in vagen Umrissen zu entwerfen, sondern diesen Entwurf auch Schritt für Schritt durchzuführen und im Einzelnen auszubauen. Dazu hätte es eines gewissen Mindestmaßes von Konzentrations- und Gestaltungskraft bedurft, das Kant nicht mehr besitzt. [. . .] Während der Arbeit ermatten auch an günstigen Tagen seine geistigen wie körperlichen Kräfte sehr rasch, und statt Begonnenes zu vollenden, beginnt er von neuem, aber nur, um wieder von neuem dasselbe Schicksal zu erleiden.' My translation.

32 Lehmann, Gerhard. 'Kants Lebenskrise' in *Beiträge zur Geschichte und Interpretation der Philosophie Kants* (Berlin: de Gruyter, 1969), 416. 'Das Nachlasswerk ist auch das entscheidende Dokument für die Alterserkrankung bzw. die sog. Senilität Kants, über die genauere pathographische Untersuchungen immer noch nicht vorliegen. Denn es ist ja hier zu unterscheiden zwischen der stets bemerkten, von Kant an zahlreichen Briefstellen beklagten, schon Anfang der 90er Jahre einsetzenden Altersschwäche, und der 1802/3 beginnenden eigentlichen Gehirnkrankheit (Dementia senilis), also der Alterspsychose.' My translation.

33 Ibid.

34 OP 22:772.

35 OP 22:788.

36 What remains to be added to this for now is that a slim tradition had already begun to write and publish medical-physiological assessments of Kant's senility from as early as 1872, as pointed out by Schepelmann. See Schepelmannn, *Der Senile Kant? Zur Wiederlegung einer populären,* 32. See also Bohn, Heinrich. 'Ueber Kant's Beziehungen zur Medizin', *APM* 9 (1872): 609–27. I would like to add Furtmayr-Schuh, Annelies. *Das große Vergessen. Die Alzheimer Krankheit* (Zürich: Kreuz Verlag, 1992), 25–6, in which Furtmayr-Schuh quotes from a private conversation with the psychiatrist Alexander Kurz "'[. . .]Der alte Herr war einer der

bedeutendsten Denker – Immanuel Kant. Die höchste Entfaltung des Verstandes und sein völliger Ruin im Alter sind niemals härter aufeinandergetroffen als in der Person des Philosophen aus Königsberg"'. Also referenced in Weinrich, *Lethe, Kunst und Kritik des Vergessens*, 103.

37 Schepelmann, *Der senile Kant?*, 9.

38 Schepelmann, *Der senile Kant?*, 161–91.

39 Adickes, *Kants Opus Postumum*, 605. 'Kant war von der Ausarbeitung seines Systems so stark in Anspruch genommen und hatte andererseits mit dem zunehmenden Alter die Fähigkeit, sich in fremde, von dem eigenen erheblich abweichende Gedankengänge hineinzufinden, so sehr verloren, daß er nicht versuchte, in die Entwicklung seiner Schule mit einem Machtwort öffentlich einzugreifen.' My translation.

40 See Schepelmann, *Der senile Kant?*, 195–207. Available at *Digitalisierte Sammlungen der Staatsbibliothek zu Berlin – Preußischer Kulturbesitz.*

41 See Schepelmann, *Der senile Kant?*, 193–4.

42 Herder, Johann G. *Werke in zehn Bänden – Frankfurter Ausgabe* (Frankfurt a. M.: Deutscher Klassiker Verlag, 1985–1998).

43 Immanuel Kants Logik, ein Handbuch zu Vorlesungen*, ed. Gottlob B. Jäsche (Königsberg: Friedrich Nicolovius, 1800),* AA 9: 1–150; Immanuel Kants physische Geographie*, ed. and in part revised at the author's request, from his own manuscript, by Friederich Theodor Rink (Königsberg: Göbbels and Unzer, 1802)*, AA 9:151–437; Über Pädagogik*, ed. Friedrich Theodor Rink (Königsberg: Friederich Nicolovius, 1803),* AA 9: 437–85.

44 Schepelmann, *Der senile Kant?*, 9; 90–102.

45 Ibid., 9.

46 Ibid., 9–10; 111–38.

47 Ibid., 74–5. The letters are printed in Vaihinger, Hans. 'Briefe aus dem Kantkreise', *APM* 17 (1880): 286–99.

48 Schepelmann, *Der senile Kant?*, 16. 'Virulente Vergesslichkeit.' My translation.

49 Ibid.

50 See, for instance, Kant's Memorierbuch (memory book 17 August until 25 September 1802), https://haab-digital.klassik-stiftung.de/viewer/image/3944917103/6/#topDocAnchor, accessed 30.05.2023.

51 Schepelmann, *Der senile Kant?*, 70–1.

52 Ibid., 29.

53 Ibid.

54 But could one not also read the scholarship on the *Opus Postumum* since Adickes as nothing but the attempt of each and every scholar to assess and judge for themselves along what line or paragraph the *Opus Postumum* is to be read either critically, and thus as an evolution or further confirmation of Kant's

Transcendental Philosophy, or clinically and thus as symptomatic of a senile, atrophying spirit who, due to his forgetfulness, ended up betraying his own philosophical principles?

55 Schepelmann, *Der senile Kant?*, 71. 'Folgendes kann man vermuten: Kant plagt zunehmend eine dem Alter geschuldete, also <senile> Vergesslichkeit, aber er kann ihr durch Hilfsmittel entgegenwirken. Sie wird sich im Alltag ungleich stärker ausgewirkt haben als auf dem Papier des Denkers. [. . . es ist angemessen] für die letzten Jahre zu unterscheiden zwischen dem schwachen Körper und vergesslichen Kopf einerseits und dem starken Willen und der Denkkraft Kants andererseits.' My translation.

56 See Heidegger, Martin. GA Band 34, *Vom Wesen der Wahrheit* (Frankfurt a.M.: Vittorio Klostermann, 1997), 11. The ancient Greek concept of *alatheia* carries the word root '*-lethe*' within it, referring to forgetting, oblivion and concealment, preceded by the negative prefix '*a-*' due to which Heidegger reads the ancient Greek notion of truth as unconcealment.

57 Weinrich, Harald. *Lethe, The Art and Critique of Forgetting*, trans. Steven Rendall (Ithaca, London: Cornell University Press, 2004), 4.

58 In his introduction to Stiegler's essay *Memory,* Mark B. Hansen defines Stiegler's notion of grammatization as the process of 'the exteriorization of memory in the form of discrete marks, traces, or *gramme* that forms the hypomnesic milieu for anamnesis.' See Hansen, Mark. B. 'Introduction', in *Critical Terms for Media Studies*, ed. W. J. T. Mitchell, Mark Hansen (Chicago: The University of Chicago Press, 2010), 66. See Auroux, Sylvain. *La revolution technologique de la grammatisation* (Liège: Mardaga, 1992). Derrida, Jacques, *Of Grammatology*, Corrected Edition, trans. Gayatri Chakravorty Spivak (Baltimore: The Johns Hopkins University Press, 1997).

59 Stiegler, *TT1,* 159.

60 Esposito, Elena. 'Die vergessenen Reste: Theorie und Praxis des blinden Flecks', in *Reste, Umgang mit einem Randphänomen*, ed. Andreas Becker, Saskia Reither, Christian Spies (Bielefeld: transcript Verlag, 2005), 14. I have already critiqued Stiegler's schematic account of the history of technical thought in the preceding Chapter 3.

61 Esposito, *Die vergessenen Reste*, 19. 'Das auf Texten basierende Gedächtnis ist viel stärker als das wesentlich auf Mündlichkeit beruhende Gedächtnis der rhetorischen Tradition, weil es viel mehr Daten verlässt (es vergisst viel mehr), aber ermöglicht, eine unvergleichbar größere Vielfalt an Situationen zu berücksichtigen (es erinnert sehr viel mehr). Es ist ein abstraktes Gedächtnis, das sich des Vergessens und nicht direkt der Erinnerungsfähigkeit bedient, um Erinnerung zu erhöhen.' My translation.

62 Ibid., 20.

63 Esposito points out that modernity transformed the form of memory as a whole, in the sense that the value of both sides, both the side of memorizing and forgetting, were enhanced. See Esposito, *Die vergessenen Reste*, 18.
64 See also Weinrich, *Lethe, The Art and Critique of Forgetting*, 11.
65 Abbt, *Vergessen und das Projekt der Aufklärung*, 68.
66 Weinrich, *Lethe: Kunst und Kritik des Vergessens*, 98.
67 Weinrich, *Lethe, The Art and Critique of Forgetting*, 74.
68 Kant, ANTH AA 7:131.
69 Ibid.
70 Abbt, *Vergessen und das Projekt der Aufklärung*, 69.
71 Ibid., Kant however also distinguished between the 'good' forgetting and forgetfulness (*obliviositas*), which is 'sometimes undeserved as with older people', in the *Anthropology from a Pragmatic Point of View* AA 7:185.
72 Häfner, Steffen. 'Immanuel Kant – ein Genie der Pedanterie und Pünktlichkeit'. *Fortschritte der Neurologie-Psychiatrie 77* (2009) 659.
73 Kant, ANTH AA 7:132.
74 Wasianski, *Immanuel Kant in seinen letzten Lebensjahren*, 234. 'Der Name Lampe muss nun völlig vergessen werden.' My translation.
75 Wasianski, *Immanuel Kant in seinen letzten Lebensjahren*, 234. 'Ein sonderbares Phänomen von Kants Schwäche war folgendes. Gewöhnlich schreibt man sich auf, was man nicht vergessen will.' My translation.
76 Kant, ANTH AA 7:185. The Cambridge University Press edition of the *Anthropology* translates the German *ersetzen* as compensate, but substitute or replace appear more appropriate in the given context.
77 Kant, ANTH AA 7:185.
78 Ibid., 7:184.
79 Cicero, Marcus Tullius. *De oratore I*, ed. Augustus S. Wilkins (London: Bloomsbury Publishing, 2003), 34, 157 and Cicero, Marcus Tullius. *De Oratore II*, ed. Augustus S. Wilkins (London: Bloomsbury Publishing, 2003), 74,299; Cicero, Marcus Tullius. *De finibus bonorum et malorum II*, ed. L. D. Reynolds (Oxford: Oxford University Press, 1998), 34, 104; Cicero, Marcus Tullius. *Academia II*, ed. Tobias Reinhardt (Oxford: Oxford University Press, 2023), 1, 2.
80 Kant, ANTH AA 7:184.
81 Ibid., 7:183.
82 Ibid.
83 Ibid., 7:184.
84 Ibid. Weinrich interprets the topic of a table, like that of the table of categories/judgments, or Linnaeus system as yet another mnemonic device. See Weinrich, *Lethe: Kunst und Kritik des Vergessens*, 98.
85 Esposito, *Die vergessenen Reste*, 20.

86 Borowski, *Leben und Charakter Immanuel Kants*, 65–6. 'In seinen ersten Magisterjahren empfahl er uns, die wir um ihn her saßen, den bis dahin etwa eingesammelten wissenschaftlichen Vorrat uns als zerteilt in verschiedene Behältnisse in unserm Kopfe zu gedenken – und dann, bei der Lesung eines Buchs oder Journals, in welchem eine neue, uns bis dahin unbekannte Idee vorkäme, immer die Frage zur Hand zu haben: in welches Fach oder Behältnis gehört dies, das du nun eben liesest, hin – wo bringst du es hin? – Hierdurch würde das Gelesene oder Neugelernte sich um desto unauslöschlicher eindrücken; wir würden, wenn uns auch die Idee selbst in der Folge entfiele, doch immer uns zurufen: Hiervon oder davon ist etwas in dieses oder jenes Behältnis reponiert – und bei einiger Anstrengung würde es sich alsdann wohl wieder ganz darstellen. Er glaubte, daß solche Rubrizierung des Neugelernten auch zu einem gehörigen Ordnen unsers Wissen viel beitrage.' My translation.

87 Lehmann, Gerhard. 'Zur Frage der Spätentwicklung Kants', in *Beiträge zur Geschichte und Interpretation der Philosophie Kants* (Berlin: de Gruyter, 1969), 392. 'Dabei handelt es sich weniger um den Zeitraum der Arbeit am *Opus Postumum* (1796–1803), als gerade darum, daß das Nachlasswerk kein "Werk" ist, sondern eine Folge von Entwürfen, in denen sich Kant seine Gedanken frei erschreibt.' My translation. Howard also writes in *The Material Literary Form of Kant's Opus Postumum*, that the manuscript of the *Opus Postumum* 'displays an accidental coherence of content and form. The philosophical content, a dynamic striving to effect a transition from metaphysics to physics, is echoed by the material form, in which the manuscript pages retain the trace of Kant's rethinking of his positions [. . .]'. See Howard, Stephen. 'The Material Literary Form of Kant's Opus postumum', *Pli* 29 (2018): 82.

88 Kant, *Philippi Logik*, 24:484. Translation by Stephen Howard in Howard, 'The Material Literary Form of Kant's Opus postumum', 69–70. See also Lehmann, *Zur Frage der Spätentwicklung Kants*, 392fn.

89 Adickes, Erich. 'Lose Blätter aus Kants Nachlass', *Kant-Studien 1* (1897): 240. '[. . .] nicht erst alles im Kopfe fertig konzipiert, Inhalt wie Darstellung, sondern höchstens den Gedankengang im Voraus feststellt, dann aber das Einzelne mit der Feder in der Hand durchdenkt.' My translation.

90 Adickes, Erich. AA XIV, 595. *'[. . .]* wie es seine Gewohnheit war, durch schriftliche Fixirung seiner Überlegungen sich über annoch dunkle Punkte Klarheit zu erschreiben.' My translation.

91 Borowski, *Leben und Charakter Immanuel Kants*, 78. 'Er machte sich zuvor im Kopfe allgemeine Entwürfe; dann bearbeitete er diese ausführlicher; schrieb, was da oder dort noch einzuschieben oder zur nähern Erläuterung anzubringen war, auf kleine Zettel, die er dann jener ersten flüchtig hingeworfenen Handschrift bloß beilegte. Nach einiger Zeit überarbeitete er das Ganze noch einmal und schrieb es dann sauber und deutlich, wie er immer schrieb, für den Buchdrucker ab. Späterhin

erst bediente er sich fremder Hände zum Abschreiben. Ungerne bemerkte er in diesen Abweichungen die etwaigen Abweichungen von seiner Orthographie.'

92 Mathieu, *Kants Opus Postumum*, 61.

93 Ibid., 62. 'Dabei ist es bemerkenswert, daß der zusammenhängende Text von Kant nie (mit einer einzigen, weiter unten angeführten Ausnahme) über die Grenzen des einzelnen Bogens, ja fast nie über die Grenze der einzelnen Seite hinausgeschrieben wurde. Wenn der Raum zu knapp wird, verkleinert sich die Schrift, und wenn überhaupt kein Raum mehr zur Verfügung steht, so daß einige Wörter auf die nächste Seite geschrieben werden müssen (wie z.B. die drei letzten Wörter von Seite 2 des Viii. Bogens des V. Konv.), dann werden diese <<Restwörter>> durch das Zeichen # mit der Seite verbunden zu der sie gehören. Im Prinzip fällt also die Einheit des Gedankens mit der vorgegebenen formalen Einheit des Papiers (eine Seite, ein Bogen) zusammen. Auf diese Weise werden die Seiten zugleich zu Bildern des Gedankenganges. Immer hat das einzelne Blatt, häufig auch die einzelne Seite die Aufgabe, den Gedankengang geschlossen aufzunehmen, und erhält so eine synoptische Funktion.' My translation.

94 See, for instance, OP 22:267, Fascicle IX, sheet VII, page 1: 'Die Eintheilung nach Principien a priori kann nicht anders als nach dem System der Categorien geschehen. Die bewegende Kräfte der Materie werden also nach ihrer Quantität, Qualität, Relation und Modalität zu betrachten sein.'

95 Stiegler, *TT, 3*, 48.

Bibliography

Adickes, Erich. 'Lose Blätter aus Kants Nachlass'. *Kant-Studien* 1, 1897.

Adickes, Erich. *Kants Opus Postumum*. Berlin: Reuther&Reichard, 1920.

Adorno, Theodor W. and Horkheimer, Max. *Dialectic of Enlightenment*, translated by John Cumming. New York: Continuum, 1989.

Agamben, Giorgio. *The Sacrament of Language: An Archeology of the Oath*, translated by Adam Kotsko. Stanford: Stanford University Press, 2011.

Aristotle. *The Metaphysics*, translated by H. Lawson-Tancred. London: Penguin Books, 2004.

Aristotle. *The Nicomachean Ethics*, translated by J. A. K. Thomson. London: Penguin Books, 2004.

Auroux, Sylvain. *La revolution technologique de la grammatisation*. Liège: Mardaga, 1992.

Bacon, Francis. *Neues Organon Teilband 1*. Lateinisch – deutsch. Hamburg: Felix Meiner Verlag, 1990.

Basile, Giovanni Pietro. *Kants 'Opus Postumum' und seine Rezeption*. Berlin: de Gruyter, 2013.

Baumgarten, Alexander. *Metaphysics. A Critical Translation with Kant's Ellucidations, Selected Notes and Related Materials*, translated by Courtney Fugate, John Hymers. London: Bloomsbury Academic Publishing, 2018.

Beck, Jacob S. *Erläuternder Auszug aus den Kritischen Schriften des Herrn Prof. Kant*. Riga: Hartknoch, 1794.

Beiser, Frederick C. *German Idealism: The Struggle against Subjectivism, 1781–1801*. Cambridge, MA: Harvard University Press, 2008.

Bermes, Christian. *Welt: Vom metaphysischen natürlichen Weltbegriff*. Hamburg: Felix Meiner Verlag, 2004.

Blumenberg, Hans. *Schriften zur Technik*. Berlin: Suhrkamp Verlag, 2015.

Bohn, Heinrich. 'Ueber Kant's Beziehungen zur Medizin'. *APM* 9 (1872).

Borowski, Ludwig E. 'Leben und Charakter Immanuel Kants'. In Borowski, Jachmann und Wasianski. *Immanuel Kant. Sein Leben in Darstellungen von Zeitgenossen*. Darmstadt: Wissenschaftliche Buchgesellschaft, 2012.

Boswell, Terry. 'On the Textual Authenticity of Kant's Logic'. *History and Philosophy of Logic* 9, no. 2 (1988) : 193–203.

Brandt, Reinhard. *Kants Vorarbeiten zum 'Übergang von der Metaphysik der Natur zur Physik'. Probleme der Edition*. In *Übergang: Untersuchungen zum Spätwerk Immanuel Kants*. Frankfurt a.M.: Vittorio Klostermann, 1991.

Brandt, Reinhard. *Kritischer Kommentar zu Kants Anthropologie in pragmatischer Hinsicht* (1798). Hamburg: Meiner, 1999.

Caygill, Howard. *Art of Judgment*. Oxford: Basil Blackwell, 1989.
Cicero, Marcus Tullius.*De finibus bonorum et malorum II*, edited by L. D. Reynolds. Oxford: Oxford University Press, 1998.
Cicero, Marcus Tullius. *De oratore I*, edited by Augustus S. Wilkins. London: Bloomsbury Publishing, 2003.
Cicero, Marcus Tullius. *De Oratore II*, edited by Augustus S. Wilkins. London: Bloomsbury Publishing, 2003.
Cicero, Marcus Tullius. *Academia II*, edited by Tobias Reinhardt. Oxford: Oxford University Press, 2023.
Combes, Muriel. *Gilbert Simondon and the Philosophy of the Transindividual*, translated by Thomas LaMarree. Cambridge: MIT Press, 2013.
Deleuze, Gilles. *What Can a Body Do?* In *Expressionism in Philosophy: Spinoza*, translated by Martin Joughin. New York: Zone Books, 1992.
Deleuze, Gilles. *Desert Islands and Other Texts 1953–1974*, translated by Michael Taormina. New York: Semiotext(e), 2004.
Derrida, Jacques. *Of Grammatology*, translated by Gayatri Chakravorty Spivak, corrected edition. Baltimore: Johns Hopkins University Press, 1997.
Diogenes Laertius. *Lives of Eminent Philosophers*. Cambridge: Cambridge University Press, 2017.
Emundts, Dina. *Kants Übergangskonzeption im Opus Postumum: zur Rolle des Nachlasswerkes für die Grundlegung der empirischen Physik*. Berlin: de Gruyter, 2004.
Espinoza, Tania, '*The Technical Object of Psychoanalysis*' in *Stiegler and Technics*, edited by Christina Howells, Gerald Moore. Edinburg: Edinburgh University Press, 2013.
Esposito, Elena. 'Die vergessenen Reste: Theorie und Praxis des blinden Flecks'. In *Reste, Umgang mit einem Randphänomen*, edited by Andreas Becker, Saskia Reither, Christian Spies. Bielefeld: transcript Verlag, 2005.
Ferrarin, Alfredo. *The Powers of Pure Reason; Kant and the Idea of Cosmic Philosophy*. Chicago: The University of Chicago Press, 2015.
Fichte, Immanuel Hermann. *Johann Gottlieb Fichte's Leben und literarischer Briefwechsel*. Leipzig: Brockhaus,1862.
Fink, Eugen. *Einleitung in die Philosophie*. Würzburg: Königshausen u Neumann, 1985.
Fischer, Kuno. *Geschichte der neuern Philosophie. Bd. 3: Entstehung und Begründung der kritischen Philosophie. Die Kritik der reinen Vernunft. Immanuel Kant. Entwicklungsgeschichte und System der kritischen Philosophie*. Mannheim: Bassermann, 1860.
Fischer, Kuno. *Das Streber- und Gründerthum in der Literatur. Vademecum für Hrn. Pastor Krause in Hamburg* in *Beilage zur Allgemeinen Zeitung (München)* Nr. 148 from 25 May 1884.
Förster, Eckhart. 'Kant's Selbstsetzungslehre'. In *Kant's Transcendental Deductions. The Three ‚Critiques' and the ‚Opus postumum'*, edited by Eckhart Förster. Stanford: Stanford University Press, 1989.

Förster, Eckhart. 'Introduction to the Opus Postumum'. In Kant, Immanuel. *Opus Postumum*, translated by Eckhart Förster, Michael Rosen. Cambridge: Cambridge University Press, 1993.

Förster, Eckhart. *Kant's Final Synthesis.* Cambridge, MA: Harvard Unversity Press, 2000.

Foucault, Michel. *'Practicing Criticism'*. In *Politics, Philosophy, Culture: Interviews and Other Writings, 1977-1984*, translated by Alan Sheridan and others. London: Routledge, 1988.

Foucault, Michel. *Introduction to Kant's Anthropology*, translated by Roberto Nigro, K. Briggs. Los Angeles: Semiotext(e) Foreign Agents Series, 2008.

Freud, Sigmund. *Totem and Taboo*, translated by J. Strachey. London: Routledge Classics, 2004.

Furtmayr-Schuh, Annelies. *Das große Vergessen. Die Alzheimer Krankheit.* Zürich: Kreuz Verlag, 1992.

Guyer, Paul. 'Reason and Reflective Judgment: Kant on the Significance of Systematicity'. *Noûs* 24, no. 1 (1990): 17–43.

Häfner, Steffen. 'Immanuel Kant – ein Genie der Pedanterie und Pünktlichkeit'. *Fortschritte der Neurologie-Psychiatrie* 77 (2009): 655–61.

Hall, Bryan. *The Post-Critical Kant: Understanding the Critical Philosophy through the Opus Postumum.* New York: Routledge, Taylor Francis Group, 2015.

Hansen, Mark. B. *Introduction* in *Critical Terms for Media Studies*, edited by W. J. T. Mitchell, Mark Hansen. Chicago: The University of Chicago Press, 2010.

Hasse, Johann G. *Merkwürdige Aeusserungen Kant's von einem seiner Tischgenossen.* Königsberg: Hering, 1804.

Heidegger, Martin. *Qu'est-ce que la métaphysique?*, translated by Henry Corbin. Paris: Gallimard, 1938.

Heidegger, Martin. *The Question Concerning Technology*, translated by William Lovitt. New York and London: Garland Publishing, 1977.

Heidegger, Martin. *Die Selbstbehauptung der deutschen Universität.* Frankfurt a. M.: Klostermann, 1983.

Heidegger, Martin. *GA Band* 34, *Vom Wesen der Wahrheit.* Frankfurt a.M.: Vittorio Klostermann, 1997.

Heidegger, Martin. *GA Band 90, Zu Ernst Jünger.* Frankfurt a.M.: Klostermann, 2004.

Heidegger, Martin. *Sein und Zeit.* Tübingen: Max Niemeyer Verlag, 2006.

Heidegger, Martin. *GA Band 76, Vorstudien zum Technik-Vortrag.* Frankfurt a.M.: Klostermann, 2009.

Heidegger, Martin. *GA Band 86: Hinweise und Aufzeichnungen, Seminare Hegel – Schelling.* Frankfurt a.M.: Klostermann, 2011.

Heidegger, Martin. *The Question Concerning the Thing*, translated by James D. Reid, Benjamin D. Crowe London: Rowman & Littlefield, 2018.

Heisenberg, Werner. *Das Naturbild der heutigen Physik* in *Die Künste im technischen Zeitalter*. Hamburg: Rohwohlt, 1956.

Herder, Johann G. *Werke in zehn Bänden – Frankfurter Ausgabe.* Frankfurt a. M.: Deutscher Klassiker Verlag, 1985–1998.

Hesiod. *Theogony*, translated by M. L. West. Oxford: Oxford Worlds Classics, 2009.

Höffe, Ottfried. *Kants Kritik der reinen Vernunft: eine Philosophie der Freiheit.* München: Beck, 2012.

Howard, Stephen. 'The Material Literary Form of Kant's Opus Postumum'. *Pli* 29 (2018): 65–87.

Howard, Stephen. 'The Transition within the Transition: the Übergang from the Selbstsetzungslehre to the ether proofs in Kant's Opus Postumum'. *Kant-Studien* 110, no. 4 (2019): 595–617.

Howard, Stephen. *Kant's Late Philosophy of Nature.* Cambridge: Cambridge University Press, 2023.

Hui, Yuk. *Recursivity and Contingency.* Washington, DC: Rowman&Littlefield Publishers, 2019.

Hui, Yuk. *The Question Concerning Technology in China.* Falmouth: Urbanomic, 2016.

Jachmann, Reinhold Bernhard. *Immanuel Kant geschildert* in *Briefen an einen Freund* in *Immanuel Kant: Sein Leben in Darstellungen von Zeitgenossen.* Darmstadt: Wissenschaftliche Buchgesellschaft, 2012.

Jünger, Ernst. *An der Zeitmauer.* Stuttgart: Klett-Cotta, 2013.

Jünger, Ernst. *The Worker: Dominion and Form*, translated by Bogdan Costea, L. P. Hemming. Evanston: Northwestern University Press, 2017.

Kant, Immanuel. *Opus Postumum:* Online Editionder Berlin-Brandenburgischen Akademie der Wissenschaften, accessed 1 June 2023, http://telota.bbaw.de/kant_op/.

Kant, Immanuel. *Critique of Pure Reason*, edited by Karl Rosenkranz. Leipzig: Voss, 1838.

Kant, Immanuel. *Gesammelte Schriften. 29. Vols.* Herausgegeben von der Preussischen Akademie der Wissenschaften (I-XXII), der deutschen Akademie der Wissenschaften zu Berlin (XXIII) und der Akademie der Wissenschaften zu Göttingen (XXIV-XXIX) (Abteilung I: Werke: I-IX; Abteilung II: Briefwechsel: X-XIII; Abteilung III: Handschriftlicher Nachlass: XIV-XXIII; Abteilung IV: Vorlesungen: XXIV-XXIX) Berlin: 1902-.

Kant, Immanuel. *Lectures on Logic*, translated by J. Michael Young. Cambridge: Cambridge University Press, 1992.

Kant, Immanuel. *Opus Postumum*, translated by Eckart Förster, Michael E. Rosen. Cambridge: Cambridge University Press, 1993.

Kant, Immanuel. *1998 Raising the Tone of Philosophy: Late Essays by Immanuel Kant, Transformative Critique by Jacques Derrida.* Baltimore: Johns Hopkins University, 1993.

Kant, Immanuel. 'Critique of Practical Reason'. In *Practical Philosophy*, translated by M. J. Gregor. Cambridge: Cambridge University Press, 1996.

Kant, Immanuel. 'Groundwork of the Metaphysics of Morals'. In *Practical Philosophy*, translated by Mary J. Gregor. Cambridge: Cambridge University Press, 1996.

Kant, Immanuel. *Critique of the Power of Judgment*, translated by Paul Guyer, Eric Matthews. Cambridge: Cambridge University Press. 2000.

Kant, Immanuel. *Prolegomena to Any Future Metaphysics with Selections from the Critique of Pure Reason*, Revised Edition, translated and edited by Gary Hatfield. Cambridge: Cambridge University Press, 2004.

Kant, Immanuel. 'The only Possible Argument in Support of a Demonstration of the Existence of God'. In *Theoretical Philosophy, 1755-1770*, translated by David Walford, Ralf Meerbote. Cambridge: Cambridge University Press, 2007.

Kant, Immanuel. *Theoretical Philosophy after 1781*, translated by G. Hatfield, M. Friedman, H. Allison, P. Heath. Cambridge: Cambridge University Press, 2010.

Kemp Smith, Norman. *A Commentary to Kant's Critique of Pure Reason.* Charleston: Forgotten Books, 2012.

Klinge, Hendrik. *Die moralische Stufenleiter. Kant über Teufel, Menschen, Engel und Gott.* Berlin: Walter de Gruyter, 2018.

Krause, Albrecht. *Immanuel Kant wider Kuno Fischer zum ersten Male mit Hülfe des verloren gewesenen Kantischen Hauptwerks, Vom Übergang der Metaphysik zur Physik' vertheidigt von Albrecht Krause. Eine Ergänzung der Populären Darstellung der Kritik der Reinen Vernunft in der Lehre vom Gegenstand und Ding an sich.* Lahr: Schauenburg, 1884.

Kühn, Manfred. *Kant. Eine Biographie.* München: Beck, 2003.

Lambert, Johann Heinrich. *Neues Organon oder Gedanken über die Erforschung und Bezeichnung des wahren und dessen Unterscheidung vom Irrtum und Schein.* Berlin, Boston: Akademie Verlag, 1990.

Laugstien, Thomas. *Philosophieverhältnisse im deutschen Faschismus.* Hamburg: Argument, 1990.

Lehmann, Gerhard. *Beiträge zur Geschichte und Interpretation der Philosophie Kants.* Berlin: Walter de Gruyter & Co., 1969.

Lehmann, Gerhard. *Kants Tugenden: Neue Beiträge zur Geschichte und Interpretation der Philosophie Kants.* Berlin: de Gruyter, 1980.

Ludwig, Bernd. *Kants Rechtslehre.* Hamburg: Meiner, 1988.

Malabou, Catherine. *Before Tomorrow: Epigenesis and Rationality*, translated by Carolyn Shread. Cambridge: Polity Press, 2016.

Mathieu, Vittorio. *Kants Opus Postumum.* Frankfurt a. M.: Klostermann, 1989.

Mathieu, Vittorio. *Erfinderische Vernunft in Kants Opus Postumum* in *Übergang: Untersuchungen zum Spätwerk Immanuel Kants.* Frankfurt a. M.: Vittorio Klostermann, 1991.

Mehring, Reinhard. 'Tradition und Revolution in der Berliner Universitätsphilosophie' in Von Bruch, Rüdiger. *Die Berliner Universität in der NS-Zeit.* Band II: Fachbereiche und Fakultäten. Stuttgart: Franz Steiner Verlag, 2005.

Micraelius, Joh. *Micraelii lexicon philosophicum..* Jenae: Impensis J. Mamphrasii, 1653.

Moss, Jessica. *Plato's Epistemology: Being and Seeming.* Oxford: Oxford University Press, 2021.

Rink, Friedrich Th. *Fragmente aus Kant's Leben: ein biographischer Versuch.* Königsberg: Hering & Haberland, 1807.

Rousseau, Jean-Jacques. 'Rêveries du promeneur solitaire'. In *Œuvres complètes*, vol. 1–4. Paris: Gallimard, 1959–95.

Sandkuhler, Hans Jorg. *Philosophie im Nationalsozialismus.* Hamburg: Felix Meiner Verlag, 2009.

Santozki, Ulrike. *Die Bedeutung antiker Theorien für die Genese und Systematik von Kants Philosophie Eine Analyse der drei Kritiken.* DeGruyter: Berlin, 2006.

Schäfer, Dieter. 'Aspekte der Wirtschaftsgeschichte Würzburgs vom Ausgang des Alten Reichs bis zur Gegenwart. Probleme, Projekte, Entwicklungen, Märkte, Betriebe, Firmen, Niederlassungen, Beschäftigung, Unternehmer und die Rolle der Stadt in zwei Jahrhunderten'. In *Geschichte der Stadt Würzburg*, edited by Ulrich Wagner. Darmstadt: Wissenschaftliche Buchgesellschaft, 2001.

Schepelmannn, Maja. *Der Senile Kant? Zur Wiederlegung einer populären These.* Paderborn: Brill, 2018.

Schopenhauer, Arthur. *Die Welt als Wille und Vorstellung, Bd 1: Vier Bücher, nebst einem Anhang, der die Kritik der Kantischen Philosophie enthält.* Leipzig: Brockhaus, 1819.

Schlosser, Johann Georg. *Plato's Briefe nebst einer historischen Einleitung und Anmerkungen.* Königsberg: Nikolovius, 1795.

Sebestik, Jan. 'The Rise of the Technological Science'. *History and Technology: An International Journal* 1 (1983).

Seibicke, Wilfried. *Technik. Versuch einer Geschichte der Wortfamilie um τέχνη in Deutschland vom 16. Jahrhundert bis etwa 1830.* Düsseldorf: VDI Verlag, 1968.

Simondon, Gilbert. *L'Individu et sa genèse physico-biologique.* Paris: Presses universitaires de France, 1964.

Simondon, Gilbert. *On the Mode of Existence of Technical Objects*, translated by Cecile Malaspina, John Rogove. Minneapolis: Univocal, 2017.

Simondon, Gilbert. *Individuation in Light of Notions of Form and Information*, translated by Taylor Adkins. Minnesota: University of Minnesota Press, 2020.

Stiegler, Bernard. *Technics and Time, 1. The Fault of Epimetheus*, translated by Richard Beardsworth, George Collins. Stanford: Stanford University Press, 1998.

Stiegler, Bernard. *Pharmacology of Spirit* in *Theory after Theory*, edited by Jane Elliott, Derek Attridge. London: Routledge, 2011.

Stiegler, Bernard. *Technics and Time, 3. Cinematic Time and the Question of Malaise*, translated by Stephen Barker. Stanford: Stanford University Press, 2011.

Tilitzki, Christian. *Die Deutsche Universitätsphilosophie in der Weimarer Republik und im Dritten Reich.* Berlin: Akademie Verlag, 2002.

Toffler, Alvin. *Science and Change* in Isabelle Stengers & Ilya Prigogine, *Order out of Chaos.* London: Verso Books, 2017.

Vaihinger, Hans. *Die Philosophie des Als-Ob. System der theoretischen, praktischen und religiösen Fiktionen der Menschheit auf Grund eines idealistischen Positivismus*. Saarbrücken: Dr Müller, 2007.

Vaihinger, Hans. 'Eine Blattversetzung in Kants Prolegomena'. *Philosophische Monatshefte* 15 (1879).

Vaihinger, Hans. 'Briefe aus dem Kantkreise'. *APM* 17 (1880).

Vaihinger, Hans. 'Zu Kants Widerlegung des Idealismus'. In *Straßburger Abhandlungen zur Philosophie*. Freiburg: Mohr, 1884.

Vernant, Jean-Pierre. *Myth and Society in Ancient Greece*, translated by Janet Lloyd. New York: Zone Books, 1990.

Vernant, Jean-Pierre. *Myth and Thought among the Greeks*, translated by Janet Lloyd, Jeff Fort. New York: Zone Books, 2006.

Wasianski, E. A Ch. *Immanuel Kant in seinen letzten Lebensjahren in Borowski, Jachmann und Wasianski. Immanuel Kant. Sein Leben in Darstellungen von Zeitgenossen*. Darmstadt: Wissenschaftliche Buchgesellschaft, 2012.

Weatherby, Leif. *Transplanting the Metaphysical Organ*. New York: Fordham University Press, 2016.

Weinrich, Harald. *Lethe, Kunst und Kritik des Vergessens*. München: C.H. Beck, 1997.

Weinrich, Harald. *Lethe, The Art and Critique of Forgetting*, translated by Steven Rendall. Ithaca, London: Cornell University Press, 2004.

Zammito, John H. *The Genesis of Kant's Critique of Judgment*. Chicago and London: University of Chicago Press, 1992.

Dictionaries

Deutsches Wörterbuch von Jacob Grimm und Wilhelm Grimm, accessed 28 May 2023, https://www.dwds.de/wb/dwb/trieb.

Dictionary of Untranslatables, edited by Barbara Cassin, translated by Steven Rendall, Christian Hubert, Jeffrey Mehlman, Nathaniel Stein, Michael Syrotinski. Princeton: Princeton University Press, 2014.

Walch, Johann G. *Philosophisches Lexicon*. Leipzig: Verlegts Joh. Friedrich Gleditschens seel. .

Wolfgang, Pfeifer. *Etymologisches Wörterbuch des Deutschen*. Digitales Wörterbuch der deutschen Sprache, 1993, accessed 29 May 2023, https://www.dwds.de/d/wb-etymwb.

Index

www.ingramcontent.com/pod-product-compliance
Ingram Content Group UK Ltd.
Pitfield, Milton Keynes, MK11 3LW, UK
UKHW020250290926
15470UKWH00002B/835

9 781350 299030